Maria do Socorro Bezerra da Silva
André Luis Lopes Moriyama
Carlson Pereira de Souza

REMOVAL OF HEAVY METALS FROM PRINTED CIRCUIT BOARDS

Maria do Socorro Bezerra da Silva
André Luis Lopes Moriyama
Carlson Pereira de Souza

REMOVAL OF HEAVY METALS FROM PRINTED CIRCUIT BOARDS

AN ELECTRODEPOSITION STUDY USING COPPER ELECTRODES

ScienciaScripts

Imprint
Any brand names and product names mentioned in this book are subject to trademark, brand or patent protection and are trademarks or registered trademarks of their respective holders. The use of brand names, product names, common names, trade names, product descriptions etc. even without a particular marking in this work is in no way to be construed to mean that such names may be regarded as unrestricted in respect of trademark and brand protection legislation and could thus be used by anyone.

Cover image: www.ingimage.com

This book is a translation from the original published under ISBN 978-620-2-55826-6.

Publisher:
Sciencia Scripts
is a trademark of
International Book Market Service Ltd., member of OmniScriptum Publishing Group
17 Meldrum Street, Beau Bassin 71504, Mauritius
Printed at: see last page
ISBN: 978-620-2-83641-8

SILVA, M. S. B. da - Study of the removal of heavy metals from printed circuit boards by electrodeposition using copper electrodes. PhD Thesis, UFRN, Graduate Program in Chemical Engineering, Area of Concentration: Chemical Engineering, Research Line: Separation Engineering and Thermodynamics, Natal, Brazil, 2018.

Advisor: Prof. Dr. Carlson Pereira de Souza.
Coorientador: Prof. Dr. André Luis Lopes Moriyama.

Abstract: E-waste is becoming one of the most serious environmental and social problems of today. To raise awareness of the problem, it is necessary to discuss disposal and recycling alternatives. In this research, a new methodology for the removal of metals from Printed Circuit Boards (PCBs) was studied. It will contribute to the minimization of environmental impacts by recovering the PCI's metallic elements, which will reduce the exploitation of mineral deposits. The process of PCI's recycling presented in this thesis, consisted in the application of technologies of chemical and physical nature, involving the leaching stage in acid medium (nitric and hydrochloric) and removal of copper, silver and tin by electrodeposition through copper electrodes. The electrodeposition reactions were performed in three current conditions (0.5 A, 1.0 A and 1.5 A) and time variations (15, 30, 60, 90 and 120 min.) respectively. For all conditions, the process was performed without mechanical agitation and with mechanical agitation of 550 rpm. In the next step, the leached solution was diluted in proportions of 1:3 and 1:7, and electrodeposition tests were performed under the same current conditions, with and without agitation and a constant time of 60 minutes. The electrodeposition of copper was shown to be efficient in all times and currents applied, removing over 98%. For silver, the best deposition rates occurred at a current density of 0.022 A/cm2 and time of 60 minutes, removing above 7%. For tin, the best removals occurred without stirring, with the solution diluted, presenting current removal rates of 1.5 A and without stirring of approximately 99%. The silver and tin concentrations at the end of each process were analyzed by X-Ray Fluorescence (XRF), Atomic Absorption Spectroscopy (EAA), X-Ray Diffraction. The morphologies of the deposits in the cathode were analyzed by Scanning Electron Microscopy (SEM) coupled with Dispersive Energy Spectroscopy.

Keywords: Recycling, Printed Circuit Boards, copper, silver, tin, electroplating.

Dedication

His disappointment, his broken heart, the traumas that remained are not the story of his life. The story of your life, are all the times that you got up after those falls, stronger, more sensitive and more courageous. The story of your life are all the times you've managed to extract the best from the worst. That's the story you have to tell. I dedicate this thesis to my parents Jacinto and Josefa, family and friends.

I know, if I got here, if I got what I wanted;

I came because I stubbornly bet on crossing;

I didn't do everything I wanted, but I'm happy, I wasn't perfect;

I made a mistake, but I tried to do it right.

I walked, I ran, I flew, I got in the way, I lost the plumb line;

I came back, I started over, I redesigned, I found my way;

I didn't do everything I wanted, but I'm happy;

No, I wasn't perfect;

I made a mistake, but I tried to do it right...

.... I made a mistake, but I ended up pure soul...

Alma Pura - Silvio Brito

Summary

Chapter 1

Introduction

1. Introduction

E-waste is becoming one of the most serious environmental and social problems of today. Behind high-tech devices, such as cell phones, computers and tablets, there is the voracious cost of degradation of the environment, water, soil, air. If you become aware of the problem, discussing disposal and recycling alternatives is necessary in the current scenario.

Brazil is among the largest producers of electronic waste in the world, with over 1.4 million tons produced annually. The electronic industry in Brazil is growing exponentially every year, developing updates for its products in a period as short as six months. Thus, the country is already among the largest producers of electronic waste in the world, totaling approximately 7 kg per inhabitant. According to the UN, Brazil is the country in the world that most disposes of outdated equipment in nature (EXAME, 2016).

Electronic equipment, besides noble metals, usually brings in its components a large amount of dangerous metals, which can cause serious impact to the environment when improperly disposed of.

Recent research shows that the impact is not only environmental but also economic, since most of these materials can be recycled and returned to the manufacturing process, saving electricity and natural resources. Recent public policies have encouraged investments in the electronics recycling market. However, additional costs reduce companies' interest in recycling this type of waste. Because they are still recent, the policies for the treatment of garbage in Brazil are not yet fully widespread among the majority of the population, only 13% of the electronic waste produced in the country is treated correctly. According to data from the Ministry of the Environment, 500 million pieces of equipment remain unused in the homes. The PNRS (National Policy for Solid Waste), Law No. 12,305, 2010, establishes that all institutions and organizations are responsible for separating and correctly disposing of the waste they produce (MMA, 2010).

In recent decades, many studies have been carried out to develop environmentally benign production processes that allow parallel recovery of metals and efficient treatment of toxic compounds. It is important to find reliable and economical heavy metal recovery techniques that do not produce secondary pollution threats to the environment and human health during processing. One of the possible solutions is provided by electrochemical processes that have high environmental compatibility, due to the fact that the main reagent, the electron, is a "clean reagent" (Janssen and Koene, 2002; Scott, 1995; Walsh, 2001). By using direct or mediated electrochemical oxidation for the dissolution of metals, reagent consumption can be significantly reduced (Arslan and Duby, 1997). In addition, in parallel with metal oxidation, electroextraction can be an alternative method to obtain the leaching solution targets(Jandova et al., 2000; Veglio et al., 2003; Veit et al., 2005).

The electrochemical technology, through the use of electrochemical reactors, offers an efficient alternative for the control of the concentration of metallic ions in aqueous solution.

The development of this research was driven by the need to minimize the environmental impact caused by the improper disposal of electronic equipment of the most diverse types and heterogeneous composition, most of them toxic. However, some of these metals have high economic value and can be extracted by safe, fast, cheap and clean techniques. The reuse of these metals is possible, since they are extracted in a selective way and in their metallic form, thus allowing the reuse in other scientific researches.

An innovative character is noted in this research in relation to other studies such as that of Veit et al., (2011), where the leaching tests in this study were performed with additional reagents, thus passing the material comminuted by two leaching processes, one with HNO3, followed by another leaching step with NaS2O3 and NH4OH, thus requiring more time and additional expenditure in the leaching process. In this thesis, the leaching step is performed in a single step with regia water (HNO3 and HCl), for a time of 6h at 80°C, which was sufficient for all metals to be solubilized, especially copper that showed 100% leaching. Still in this work cited, for the recovery of copper, the pH of the solution was adjusted, that is, other chemical reagents are necessary to obtain copper. In this thesis, this step was not necessary, and even so, 98% of the copper is recovered in the original solution, presenting high removal rates in all the studied times (between 15 and 120 minutes), and the best removal time in 60 minutes can be determined. The mentioned work finds the best copper removal rate in 180 minutes, increasing the response time of the process, as additional expenses. In the process of this thesis there are fewer steps to obtain the copper metal, and in this same process it is possible to obtain silver, tin, as well as the removal of lead without any additional chemical reagent, besides using low cost electrodes, as is the case of copper.

In this work, we studied the obsolete Printed Circuit Boards (PCI's) of desktop computers from the Federal University of Rio Grande do Norte. The present study contemplated the removal of copper, tin and silver, among other metals, present in PCB leaching liquors by electrodeposition processes using copper electrodes, besides the use of a hydrometallurgical route that allowed the solubilization of metals in a leaching medium (aqua regia), in order to optimize their recovery in valuable and economical forms.

In this work, we tried to offer some contribution to the understanding of the physical and chemical mechanisms involved in the electrodeposition of metals. The work is organized in five chapters. In chapter 1, there is an introduction on the subject, and in chapter 2, a literature review where concepts and theories were gathered that allowed us to advance in the understanding of the experimental results described in chapter 4. The experimental methodology used is presented in chapter 3. And, in chapter 5, we present our conclusions.

1.1. General objective

Study the removal of metals from printed circuit boards, by hydrometallurgy followed by electrodeposition, using copper electrodes.

1.1.1. Specific objectives

- Design an electrochemical bench cell for the initial purpose of removing metals in solution;
- Make copper electrodes, from motor winding wires, to be applied in the electrochemical process;
- Dissolve the PCI's powders by hydrometallurgy, performing leaching with regia water;
- Investigate the influence of parameters such as current density, process time, with and without agitation, on the removal of metals;
- Analyze the electrodeposition process in solutions diluted with distilled water;
- Check the influence of the geometry of the electrodes, associated with the removal of metals;
- Assess the best metal deposition time as well as the most efficient current;
- Analyze the loss and gain of mass of the electrodes during the electrodeposition process.

Chapter II

Literature review

2   Literature review

2.2 Waste Electrical and Electronic Equipment (WEEE)

E-waste is one of the fastest growing and most challenging solid waste streams in the world due to its complex mix of metals, plastics, ceramics and hazardous components, but at the same time contains a wealth of metals embedded in its composition (SHOKRI, 2017).

Considering alternatives for the disposal of these materials in a way that no longer causes damage to the environment attracts the attention of many researchers in different fields. In addition, there is also social pressure for environmentally friendly products and systems, which in turn has led to new environmental legislation, particularly in Europe. New regulations make producers responsible for the costs of collection, treatment and recovery of their products. These rules insist that products must be designed to reduce their environmental impact, particularly with increased recycling rates. Therefore, today the task of designing a sustainable product is indispensable for society worldwide (SHOKRI, 2017).

Electrical and electronic waste is known as any peripheral component and electronic equipment such as refrigerators, electric ovens, television sets, DVD players, mobile phone devices, computers and spare parts of these equipment that are no longer useful in their original form (WIDMER et al., 2005, LI et al., 2004).

Estimates show an increase in the global production of electrical and electronic waste, with China, Eastern Europe and Latin America becoming the largest producers of electronic waste in the coming decades. To make the process of recycling electronic waste feasible, it is important to know the materials that compose them in order to target technologies to achieve the recovery of these materials. Another way is to separate electronic waste by type of equipment to determine the composition of electronic waste in ceramic, polymeric and metallic materials of specific equipment (MANETTI et al., 1996).

Among the electrical and electronic waste, one of the main components for which an effective solution for reuse and recycling has not yet been found is the printed circuit board (PCB). One of the main difficulties pointed out by several researchers is the separation of components and materials, due to their diversity, in order to perform the necessary functions in the devices. To separate the electronic components and reuse the materials, it is necessary to remove them from the solder, which is a complex process, and often makes the components unusable due to the temperature applied, which is often very high depending on the process used (SHOKRI, 2017).

Brazil also follows the worldwide trend of fast growth in the consumption of electro-electronic equipment, leading to an increase in the generation of this waste, which despite being urban, in relation to its generation, has characteristics of an industrial waste, due to its composition.

The final destination of waste electrical and electronic equipment is incineration or landfill, but when landfilled, metals such as lead, cadmium and mercury can be leached contaminating the soil. From this perspective, the recycling of waste electrical and electronic equipment contributes to increase the useful life of landfills and conservation of natural resources (SIMONI et al., 2015).

In Brazil, although the law that deals with solid waste is already in force, there are still no companies focused on recycling the metals present there, what actually happens is the collection, disassembly and segregation of waste, and the printed circuit boards, which concentrate metals of economic interest such as copper, gold and silver, are sent to recycling plants abroad, leaving only the waste of lower added value, absorbed by the domestic recycling market.

Additionally, other factors that reflect the importance of the residue of electro-electronic equipment as an alternative source for metal extraction can be highlighted, such as: the depletion of high content ores, energy savings in the recovery process since the metal is already in its metallic form, and the exploration of mineral reserves to meet the demand for raw material from the increasing amount of electro-electronic products.

2.2.1 Printed circuit board (PCB)

Printed circuit board (PCB) is the integral component of any electronic equipment as it switches on electrically and mechanically supports the other electronic components. The basic structure of PCBs is copper coated laminate consisting of glass reinforced epoxy resin and a variety of metallic materials, including precious metals. The concentration of precious metals, especially Au, Ag, Pd and Pt, is much higher than their primary resources, making PCI waste an economically attractive urban ore for recycling. In addition, PCI also contains different hazardous elements, including heavy metals, flame retardants that pose a serious hazard to the ecosystem during conventional landfill and incineration waste treatment. Consequently, electronic waste, incluindo o desperdício de PCI, são mantidos nas lojas ou enviados para os países em desenvolvimento onde o estrato mais pobre da população encontra benefícios econômicos recuperando metais preciosos em métodos rudimentares (CHI et al., 2011).

Manufacturers, environmental agencies and governments around the world are looking for systematic and environmentally friendly recycling technologies for these waste PCIs, through which all metallic and non-metallic values can be recovered, while having minimal impact on the environment. Although some articles (HUANG et al., 2009; LI et al., 2004; LUDA, 2011; SOHAILI et al., 2013) have analysed the potential aspects of different routes, however, all issues regarding the total recycling of PCI waste have not yet been addressed.

Following this approach, the objective of this thesis was to provide a study of PCI recycling, using the electrochemical process and reporting in the discussions its advantages and disadvantages to achieve a cleaner process of reuse of these wastes, with special attention to the extraction of metallic values as well as toxic metals.

2.2.2. Chemical composition

Electrical and electronic waste has more than 1000 different substances, many of them toxic, such as mercury, lead, arsenic, cadmium, selenium, hexavalent chromium and flame retardants that generate dioxin emissions when burned. The elemental composition of PCIs varies according to their type and applications. In general, they contain ~28% metals, 23% plastics and the remaining percentage as ceramics and glass materials (ZHOU e QIU, 2010). The main reinforcement material for PCI's substrate is fabric, made of glass fibers or silica. Other inorganic materials such as alumina, alkaline and alkaline-earth oxides and small amounts of other mixed oxides such as barium titanate are also present (SUM, 1991). Ceramic materials such as beryllium oxide (BeO) can also be found in PCI's.

About 10-20% of PCI is composed of copper, which forms the conductive layer for electrical connection between different components. Precious metals, especially gold (Au) and palladium (Pd), are used as contact materials in the joints. Typical Pb/Sn welds, which are used to join different PCI components, represent 4-6% of the total weight of these residues. The components that are assembled on PCI also contain different metallic values, such as gallium (Ga), indium (In), titanium (Ti), silicon (Si), germanium (Ge), arsenic (As), antimony (Sb), selenium (Se), tellurium (Te), tantalum (Ta), etc.

Thin tin or silver films are used in printed circuit boards to protect against oxidation (VEIT et al., 2005). The basic metals found mainly in printed circuit boards are used for their conductive properties.

The metal fraction consists mainly of copper ~ 16%, lead ~ 4%, iron and ferrite ~ 3%, nickel ~ 2%, silver ~ 0.05%, gold ~ 0.03%, palladium ~ 0.01%, and so on (VEIT et al., 2002; GOOSEY and KELLNER, 2003), and even rare elements such as Tantalum are covered or mixed with various types of plastics and ceramics (HOFFMANN, 1992). It is clear that electronic waste varies considerably with age, origin and manufacturer; so there is no standard composition of scrap, which in fact will also be different from the composition of PCI's used in this study.

2.3. Recycling of printed circuit boards

Due to the diversity of materials present in its composition, the most complex component of computers, or any electronic equipment, in terms of recycling, is PCI. If on the one hand this diversity of materials makes the recycling process very difficult, on the other hand, the presence of precious metals or high economic value such as copper, silver and gold, among others, makes PCI's an interesting raw material for recycling (PETRANIKOVA, 2009). The generation of harmful emissions during processing is another major barrier to recycling electronic waste (BALDÉ et al., 2015).

The main processes used in PCI recycling are mechanical processing, hydrometallurgy, pyrometallurgy and electrometallurgy.

2.3.1. Mechanical processing

In primary metallurgy, mechanical processing has been used as part of the treatment and beneficiation stages of metals and ores. Regarding the recycling of metals from electronic scrap, in general, mechanical processing has been used as a pre-treatment for metal reuse (VEIT, et al., 2002). Mechanical processing includes comminution, classification and separation (by difference of density, weight, granulometry, magnetic properties and electrical property) (HAYES, 1993).

Several authors such as (CUI and FORSSBERG, 2003) presented a review on the use of mechanical recycling of scrap from electrical and electronic equipment. (ZANG and FORSSBERG, 1998), (TENÓRIO et al., 1997), (VEIT et al., 2002), (DUAN et al., 2009), (LEE et al., 2009), among others, used mechanical processing in their work, obtaining good results.

2.3.2. Hydrometallurgy

Hydrometallurgy consists of a series of attacks of acid or caustic solutions (leaching with cyanide, with halide, with thiourea or with thiosulphate and aqua regia) with the aim of dissolving a metallic material. The solutions are then submitted to separation and purification procedures, such as precipitation of impurities, solvent or liquid-liquid extraction, adsorption, ion exchange to isolate and concentrate the metals of interest. Afterwards, the solutions are treated by electrolytic process, chemical reduction or crystallization for the recovery of metals. The focus of our study is extraction and recycling of these metals by electrochemical methods after going through mechanical and hydrometallurgical processing.

Sum (1991) addresses the main advantages of hydrometallurgical processing of electronic scrap over pyrometallurgical methods: the best environmental protection in relation to air pollution risks; easier separation of the main scrap components; lower costs (low energy consumption and recycling of chemical reagents).

Disadvantages of this type of processing are cited: the difficulty in accepting more complex electronic scrap; the need to use pre-treatment (mechanical treatment) to reduce the volume of scrap. The chemical attack is only effective if the metal is exposed, large volume of solutions necessary for the dissolution of the scrap, generation of effluents containing base metals that are corrosive, toxic or both and the generation of solid waste.

Despite some criticisms, mainly regarding the issue of the large amount of effluents generated, hydrometallurgy has been used relatively successfully, especially in copper recovery (VEIT, 2005).

Lee et al. (2003a) studied the recovery of metals and the regeneration of used solutions. Frias et al. (2004) also studied means to regenerate solutions used in the manufacture of Printed Circuit Boards and later recover metals, especially copper. The regeneration was based on electrometallurgical techniques, while the recovery of copper was done through hydrometallurgical techniques, among them solvent extraction.

Kinoshita et al. (2003) used hydrometallurgical processes to recover metals from unassembled Printed Circuit Boards, i.e., burrs or scrap from the industrial process, obtaining a nickel-rich solution, with a small amount of copper, and another copper-rich solution, practically free of impurities.

Kim et al. (2008) studied the effects of copper ions on the dissolution of copper from scrap printed circuit boards using hydrochloric acid.

Havlik et al. (2010) studied the extraction (dissolution) of copper and PCI tin after a heat treatment. According to them, the extraction of copper from pyrolized samples was higher than the extraction from non-pyrolized samples, while the amount of tin extracted from a pyrolized sample was lower than the extraction from non-pyrolized samples.

Regarding the recovery of precious metals by hydrometallurgical means, the works of Gluszczyszyn et al. (1990), Vejnar and Hrabak (1990), La Marca et al. (2002), Park and Fray (2009), among others, can be mentioned.

As one of the processes of this work, hydrometallurgy will be addressed in detail in a next topic and methodology in a practical way.

2.3.3. Pyrometallurgy

According to Hoffmann (1992), conventional pyrometallurgical processing is essentially a mechanism of metal concentration in a metallic phase and the rejection of most foreign materials in a slag phase. Pyrometallurgical processes include: incineration, melting, pyrolysis, sintering, high temperature gas phase reactions, among others.

For Sum (1991), incineration is the most common way to dispose of polymeric and other organic materials in metal concentrates. The crushed scrap can be burned in a furnace to destroy the polymers, leaving a metallic residue. The melting of a raw metal concentrate can produce impure metal alloys. These alloys can then be electrolytically refined or pyrometallurgically refined. It should be noted that in this case the thermal energy released from these materials is not recovered.

Pyrolysis is the chemical decomposition of organic materials by heating in an atmosphere with little or no oxygen. Pyrolysis of organic materials contained in PCI waste leads to the formation of oils and gases that can be used as chemical raw materials or fuels (SUM, 1991).

Pyrometallurgical processes are the most used by recycling industries for the recovery of precious metals, due mainly to some advantages presented by the process such as: accepting any type of electronic scrap, not requiring pre-treatment and having few steps (VEIT, 2005).

The main problems related to the pyrometallurgical processing of electronic scrap are the gaseous emissions which, when not controlled, can generate serious environmental and health problems due to the formation of dioxins and furans from polymers and other insulating materials. Other disadvantages of the method are: loss of metals due to volatilization; increased amount of slag due to the presence of components such as glass and ceramics, causing an increase in the loss of noble metals and base metals; low recovery rate of some metals (e.g. Sn and Pb) or impossibility of recovery of others (e.g. Al and Zn) (VEIT, 2005).

According to (GUO et al., 2009), incineration is not the best method for the treatment of materials such as those present in PCIs, as they contain inorganic materials, such as fiberglass, which can reduce the energy efficiency of burning.

De Marco et al. (2008) studied the possibility of using pyrolysis for the recovery of Waste Electrical and Electronic Equipment (WEEE). The materials were pyrolyzed in nitrogen atmosphere, in an autoclave, at 500 °C for 30 minutes. According to the authors, the metals, free of polymers, can be more easily separated and recycled, while the gases can be sources of energy to sustain the process and the liquids can be used as energy source or in chemicals.

Zhou et al. (2010) studied the use of pyrolysis in the treatment of PCI waste, obtaining good results. The process used by the authors to separate and recover PCI materials consisted of two steps, vacuum pyrolysis and vacuum centrifugal separation. Hall and Williams (2007) studied the recycling and recovery of PCB scrap materials using only the pyrolysis stage.

Both the work of Cui and Zhang (2008) and that of Huang et al. (2009) state that pyrometallurgical transformation is the most traditional technology for the recovery of precious metals from electronic scrap. However, they report that state-of-the-art plants require large investments, mainly due to the necessary environmental care, which in recent years has turned the attention to other types of processes.

2.3.4. Electrometallurgy

Electrometallurgy is the process of obtaining metals through electrolysis. These processes are usually performed in aqueous electrolytes or molten salts (VEIT, 2005).

There are several studies based on electrochemical techniques with the objective of recovering metals from various types of waste. García-Gabaldón (2006) studied the performance of an electrochemical reactor of two compartments separated in batches by a ceramic membrane in the removal of tin solutions from galvanic operations. Juarez and Dutra (2000) used linear scanning voltammetry and chronopotentiometry in the study of gold recovery from thiourea solutions by electrodeposition.

Fornari and Abbruzzese (1999) studied the selective recovery of copper and nickel from galvanic and electronic solutions using electrodeposition. Copper deposition was performed under acid conditions and nickel deposition under basic conditions.

Scott et al. (1997) in their study on the recycling of metals from pickling solutions used in the manufacture of PCI's, analyzed two recycling processes: the electrochemical recycling of metals and the combination of electrochemical deposition of copper, with the precipitation of tin and lead that could be recovered by pyrometallurgy.

Doulakas et al. (2000) studied the recovery of copper, lead, cadmium and zinc from a synthetic solution by electrodeposition and observed that, using favorable potentiostatic conditions, selective electrodeposition of the four metals was possible, with purity levels above 99%.

According to Sum (1991), if the metals are concentrated through hydrometallurgy (e.g. selective dissolution, ion exchange or solvent extraction) they can be electrodeposited directly from the aqueous solutions onto the cathode. For example, in a sulfate solution containing copper and nickel, the metal can be electrodeposited leaving the nickel in solution (SUM, 1991).

Regarding the recovery of precious metals Sum (1991) cites the following advantages of electrometallurgical processes:
- ✓ A few steps;
- ✓ The concentrate of precious metals from electrolysis represents 95 - 97% of the metal found in scrap metal;
- ✓ The amount of precious metals in the anodic sludge after casting and electrolysis as a refining step is very low;

✓ It is applicable to all types of scrap containing a surface layer of precious metals on a base metal substrate;

✓ All precious metals can be dissolved simultaneously or selectively (if necessary), and the copper-based substrate remains unchanged;

✓ The electrolyte can be recycled.

The main limitation of this technique is the need to have a pre-classified scrap.

2.4. Solid Waste Legislation in Brazil

It can be said that eleven years ago a new era began in the regulatory frameworks related to waste management in Brazil. The change began in 2007 with the enactment of the National Basic Sanitation Policy (PNSB) through Law No. 11,445. That, besides establishing guidelines for basic sanitation, the referred regulation required the inclusion of solid waste by the municipalities in their Sanitation plans (BRAZIL, 2007). Three years later, and after 20 years of proceedings in Congress, the National Policy on Solid Waste (PNRS), Law No. 12,305, of August 2, 2010 (BRAZIL, 2010), was approved.

The law provides that waste must be reused or recycled, or it must be recovered and used for energy, or it must be composted before the final disposal of the waste in landfills in order to minimize environmental impacts. For this, management tools must be used throughout the solid waste chain, i.e., from the collection, transportation, transshipment, treatment stage to final disposal (BRAZIL, 2010).

Thus, one of the objectives of the National Waste Policy is to oblige the state governments to create and structure solid waste management plans in order to prioritize the reduction in waste generation, reuse, recycling, treatment and finally the final disposal of the waste according to the rules in force.

2.5. Hydrometallurgy

The processing of the metallic residues by hydrometallurgy implies in the initial phase of this study, with the solubilization of the metals by leaching and is presented in more detail in the next points.

2.5.1. Leaching

Leaching is a hydrometallurgical operation in which the metals to be recovered are dissolved in an aqueous medium by the action of leaching agents belonging to several chemical families, the most common being acidic, alkaline and complexing agents. The solutions obtained are then subjected to separation processes, such as solvent extraction, precipitation, cementation, or ion exchange, in order to isolate and concentrate the metals of interest, and these are recovered in recoverable forms (ZHAN et al., 2014). The recovery of heavy metals from aqueous solutions is of extreme importance in various industrial and environmental frameworks, from mining to wastewater treatment and environmental remediation to waste recycling (ZHAN et al., 2014).

Hydrometallurgical processes have been recognized as an efficient method to selectively recover precious metals from typical mining fluids and decommissioned electronic components (CAYUMIL et al., 2014). There are numerous hydrometallurgical techniques that can be used. Figure 2.1 shows a flow chart with the hydrometallurgical techniques that can be applied to the treatment of electrical and electronic waste (PANT, 2012).

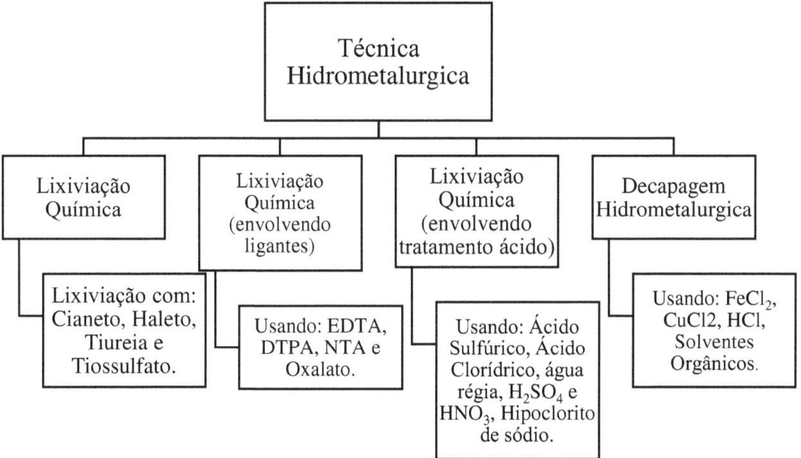

Figure 2.1 - Flowchart of hydrometallurgical techniques for treatment of WEEE (adapted from Pant et. al., 2012).

The residues of the printed circuit boards, before going through a hydrometallurgical process, are first submitted to the operations of fragmentation, classification and sampling, where the purpose of the fragmentation is to reduce the size of the material. Besides the granulometry of the material, the remaining parameters of the process (temperature, leaching agent concentration, agitation speed, liquid/solid ratio and leaching time) may also have an influence on the leaching performance of PCI, being necessary for its optimization to maximize the recovery of metals under attractive technical-economic conditions (CHOI et al, 2004).

The choice of the leaching agent depends on several factors, such as its reactivity characteristics in relation to the physical-chemical properties of the material to be leached, the selectivity, the cost of the reagent and its recycling capacity. Therefore, it was intended to choose the leaching agent nitric acid/chloric acid (aqua regia), taking into account all these items and using the existing know-how complemented by bibliographic information for the leaching agent in this study.

The use of aqua regia for the dissolution of PCI's is based on the fact that most of the elements are in metallic forms, and their solubilization will be very dependent on the existence of oxidizing conditions that allow the formation of metallic ions in environments where they are soluble. Nitric acid was one of the leaching agents chosen because it is simultaneously acidic and oxidizing, allowing the dissolution of most basic metals. Hydrochloric acid, which does not have an oxidizing character, was chosen because it is a more selective leaching agent, allowing the understanding of the leaching behavior of certain metals. If there are metals that may occur in already oxidized forms, the use of HCl allows their solubilization. Although potentially less efficient than nitric acid, hydrochloric acid promotes the oxidation of metallic phases through the H+ ion which is reduced to H2 (CHOI et al., 2004). It is known that these acids, mainly inorganic, can chemically attack copper as well as metallic silver and other metals. The high acid concentrations provide an adequate amount of hydroxide ions and continuous stirring of the solid/liquid mixture can improve incorporation and dissolution in the system.

The reaction between HNO3 and copper presents a loss of mass during acid leaching, characterized by the evolution of a brown gas. Based on the chemical reaction between metallic copper and nitric acid shown in Equation (1) a (PARK and FRAY; 2009), the brown gas is probably nitrogen dioxide (NO2).

$$Cu + 4HNO_3 \rightarrow Cu(NO_3)_2 + 2NO_2 + 2H_2O \qquad \text{Equation (1)}$$

The reaction of dissolution of silver by nitric acid is given by:

$$3Ag_{(s)} + 4HNO3_{(l)} \rightarrow 3AgNO3_{(s)} + NO_{(g)} + 2\ H2O_{(l)} \qquad \text{Equation}$$

$$(2)$$

A second reaction occurs in the presence of hydrochloric acid, which is present in royal water;

$$AgNO3_{(s)} + HCl_{(l)} \rightarrow AgCl_{(s)\ +}\ H+ + NO3\text{-}_{(l)} \qquad \text{Equation (3)}$$

Equation (3) shows that silver corrosion is due to the formation of AgNO3, which is dissolved in its form of AgCl and H+. The joint action of both the nitric acid and the complexing agent is necessary because each one has a specific function in this process. Nitric acid is a powerful oxidizer that will dissolve small amounts of silver and other metals, forming silver ions (Ag+). Hydrochloric acid is an excellent example of a source of chloride ions (Cl-), which work as complexing agents. These chloride ions will react with silver to produce anions (AgCl-). The reaction of metallic silver, considering a system with nitric acid and concentrated hydrochloric acid, can be described as shown in Equations (4) and (5).

$$Ag_{(s)} + 2\ HNO3_{(aq)} \rightarrow Ag+_{(aq)\ +}\ NO3\text{-}\ _{(aq)} + NO2_{(g)\ +} H2O_{(l)} \qquad \text{Equation (4)}$$

$$Ag+_{(aq)} + HCl_{(aq)} \rightarrow AgCl_{(s)} + H+_{(aq)} \qquad \text{Equation (5)}$$

The phenomena involved in the leaching process can be dissolution, desorption, complexation and mass transport processes and can be affected by several factors such as: chemical and physical reactions between the solid/leaching agent components (VOLSKY, 1978).

Of all these factors, the concentration of the leachate and the temperature of the process are almost always very important, because the occurrence (or not) of the chemical reaction on the surface of the solid material depends on them. The liquid/solid ratio is generally used as a way to control the concentrations that are intended to be obtained in the solution resulting from the leaching, i.e., low L/S values allow obtaining more concentrated liqueurs, thus contributing to reduce the costs of processing the liqueurs. The agitation speed and the granulometry of the material have implications in terms of physical phenomena, that is, the agitation conditions the diffusion phenomena of leaching agents in the solid material and the granulometry of the material limits the specific area of the solid that is available for the occurrence of the reaction. Finally, the reaction time variable has implications for the recovery of the elements and is directly related to the speed with which the reactive processes can occur (OLIVEIRA, 2012).

Among the leaching agents used in the PCI recycling process, acids are the most widely used and their concentration depends on which metal you want to solubilize. In general, basic metals are solubilized in acid mediums that are not very concentrated, a medium in which noble metals do not dissolve (OLIVEIRA, 2012).

The solubilisation of copper, for example, can occur both in nitric and sulphuric or hydrochloric media, with or without the addition of an oxidising agent. Other metals also dissolve, totally or partially, in the presence of these acids (such as nickel, aluminium, tin, lead and silver). However, some metals show low solubility in certain media, e.g. lead in the presence of sulphuric acid and silver in nitric acid medium (OLIVEIRA, 2012). Kaya, (2016) believes that the hydrometallurgical route will be a key piece in metal recovery for future research. But according to him, developing any new clean technology to recycle valuable PCI resources will be of great importance.

2.5.2. Behaviour of metals in solution

The thermodynamics of aqueous systems is represented in the form of Pourbaix or potential/pH (Eh/pH) diagrams. These diagrams allow to visualize the various possibilities of reactions that occur in aqueous media, but not the speed at which they occur.

An example of this is reaction ($Cu \rightarrow Cu2+$) only occurs at very high levels of acidity, data from the Pourbaix diagram. This fact is typical of noble metals, not very reactive in acid environment. The possibility of the reaction to generate the cuprous ion ($Cu+$) would supposedly occur more easily, but it is known that this ion is not very stable and therefore the dissolution of copper should be based on the reaction up to the form of cupric ion ($Cu+2$). The behavior of silver is somewhat similar to that of copper. With other metals, solubilization in acid medium is more favorable. As for $Zn \rightarrow Zn2+$, $Pb \rightarrow Pb2+$, $Sn \rightarrow Sn2+$, $Ni \rightarrow Ni2+$, among others (OLIVEIRA, 2012).

For Sn (IV) and Sn (II) tin, the oxidized form is only soluble at high acidity, with $SnO2$ oxide forming relatively easily at pH just above zero. The reduced form is soluble up to higher pH values. Thus, for tin, the leaching in oxidizing medium will only result with a large excess of acid.

2.6. Electrochemical treatments

Electrochemistry has proven to be a viable alternative in solving environmental problems. Electrochemical techniques have been widely developed in the remediation of liquid and solid waste. The main advantage of this technology is its environmental compatibility, due to the fact that it does not involve the use of chemical reagents. Other advantages are related to its versatility, high energy efficiency, use at low temperatures, ease of automation and safety (KIM et al., 2002; MARTINEZ-HUITLE et al., 2009).

The first research in the field of electrochemistry dates back to the 18th century, when Alexander Volta and later Michael Faraday realized that there was a relationship between chemical reactions and electrical phenomena (CUNHA, 2010).

The object of the study of electrochemistry are the chemical phenomena involving the separation of electrical charges. In many cases this separation of charges leads to the transfer of charges on the surface of an electrode. Generally the load transfer reactions occur on different electrodes, physically separated and immersed in an ionic or electrolyte conductor. Typically, electrodes can be formed by solid metals (Pt, Al, Cu,Ti), liquid metals (Hg, amalgams), carbon and semiconductors (Si) (BARD, 2000).

In this process, in general, the chemical composition of the metal remains unchanged, however, the passage of electric current through the conductor may be accompanied by temperature variation and evolution of magnetic effects. The metallic conductors are usually used as terminals for the input and output of electric current in the electrochemical system. This function gives metals (or any other substance that presents electronic conductance) the name of *electrode*. In ionic conductors, the charge carriers have atomic or molecular dimension and have both negative and positive charges (CUNHA, 2010).

2.6.1. Electrochemistry and Electrolytic Systems

Electrochemical cells can be classified into two types: Galvanic and Electrolytic cells. When the cell spontaneously converts the energy obtained in a chemical reaction into electrical current, it has a galvanic cell. Whenever it is necessary to supply electrical energy (electrical potential), using a voltage source external to the cell to direct the reactions in the electrodes, converting electrical energy into chemical energy, we have an electrolytic cell.

In this cell, typically composed of at least two electrodes, electrolysis of the electroactive species occurs. Such cells are used industrially in the electrolysis of metals in galvanoplasty, which consists in the covering of metallic objects by other metals, also in the purification of metals and electrosynthesis.

According to GAMBOA *et al.*, (1998) electrodeposition can occur in two different ways: in potentiostatic (constant potential) or galvanostatic (constant current) mode. The potentiostatic mode is more precise, because it is possible to apply the exact reduction potential to the working electrode obtaining good quality films. In this case, a potentiostat and an experimental assembly with three electrodes (cathode, anode and reference electrode) are required. The galvanostatic mode can establish a current between the working electrode (cathode) and the counter electrode (anode) using only a simple current source. In this work, only the galvanostatic technique will be part of the experiments as we will see later.

The electrochemical cell used in this treatment system is composed of two electrodes, called anode and cathode. Depending on the material used, processes to eliminate pollutants (organic or inorganic) will occur specifically, with different efficiencies, depending on the operating conditions (current density, electrical potential, pH, temperature, concentration of pollutants, composition, conductivity of the effluent matrix, etc.) and the electrochemical reactor used.

The main parameter used to control the electrochemical process is the current density (j), that is, the current intensity used per unit area (j=A/cm2). This parameter allows a line of investigation on the influence of the amount of electrical charge applied to the system, which effectively favors the oxidation of contaminant in solution. The value of the current density can be increased to the point where it begins to trigger parallel reactions that compete with the objective of electrochemical treatment. These competitive reactions are called oxygen evolution (at the anode) and hydrogen reduction (at the cathode). The entire study involving electrochemical systems consists in identifying the most efficient operating conditions for the chosen treatment (KIM et al., 2002; MARTINEZ-HUITLE et al., 2009).

In an electrochemical cell, there are the reduction and oxidation semirreactions: in the cathode (working electrode) there is the reduction semirreaction, in which the electrode supplies electrons to a positive ion or cation. When the cation is a metallic ion, the reduction is accompanied by its deposition on the cathode, which acts as substrate. Thus, in the electrodeposition, the metal is electrodeposited according to the reaction:

$$Mn+ + ne- \rightarrow M \qquad\qquad \text{Equation (6)}$$

However, during the electrodeposition of metals in aqueous media, the secondary production of hydrogen due to water electrolysis should receive special attention. Due to the generally acid character (low pH values, resulting in abundance of H^+ in the solution) of the electrolytes used, and the proximity of the standard reduction potentials for metals and hydrogen, the participation of hydrogen in the process of electrodeposition of metals occurs in most cases (KIM et al., 2002; MARTINEZ-HUITLE et al., 2009). The release of hydrogen gas (H2) may occur according to one of the reactions:

$$2H+ + 2e-_{\rightarrow H2} \qquad\qquad \text{Equation (7)}$$
$$2H2O_{+} ne-\rightarrow H2 + OH- \qquad\qquad \text{Equation (8)}$$

Hydrogen gas tends to promote the formation of bubbles in the working electrode that may be incorporated into the film generating porosity or simply generating holes (crater-shaped) or elongated grooves in the electrodeposites on the working electrode. Normally, during a cyclic inspection voltammetry, the release of hydrogen is accompanied by a rapid increase in cathode current. Hydrogen can interfere with the electrodeposition process of the metal not only by gas formation. It can bind to the metal according to the reaction:

$$H+ + M +{}^{e-\rightarrow} M - H \quad \text{Equation (9)}$$

Depending on the metal-hydrogen (M-H) binding energy and the diffusion coefficient of hydrogen in the metal, the formation of a hydride, hydroxide or oxy-hydroxide may occur and, eventually, the subsequent reaction:

$$M - H + M - H \rightarrow 2M + H2 \quad \text{Equation (10)}$$

For any of these possibilities, there is the compromise of the morphology and structure of the film. On the other hand, in the anode, there is the semirreaction of oxidation. When the anion is a hydroxide ion, oxygen gas is produced according to the reaction:

$$4OH- \rightarrow O2 + 2H2O + 4e- \qquad\qquad \text{Equation (11)}$$

This semirreaction happens when working with inert electrodes (EC). In this case, the CE only closes the electric circuit for current circulation, without interfering in the process of reduction and deposition in the working electrode (ET).

The electrochemical reactor used in this study is batch. In a batch process all the reagents are placed inside the reactor and mixed, remaining there throughout the electrolysis process. The resulting solution is then removed from the reactor and the products are separated and purified. During the process conditions vary with time (concentration of reagents and products, temperature of the solution), requiring operational adjustments such as voltage to maximize current efficiency, however, the solution contained in the reactor is uniform, i.e. has the same composition (BORIN, 1986).

Batch reactors are often used on a laboratory scale and are unsuitable for high production rates due to the intermittent nature of their operation. They are also called batch operated reactors because there is a time-dependent functional parameter (BORIN, 1986).

These reactors, considered simple, are useful only in processes where production rates are low. For higher rates and continuous processes, more complex reactors are required, which can be fed continuously with the electrolytic solution (BORIN, 1986).

2.6.2. Distribution of current in the reactor

In order to increase the spatial time yield of electrochemical reactors, it is not enough just to increase the electrode area, it is also necessary that the current density is equally distributed throughout the electrode, so that the reaction of interest occurs with a uniform speed. However, the current distribution is only uniform for very simple geometry electrodes. In general, we should consider that the density distribution is not uniform (BORIN, 1986).

The distribution of the current density by the electrode is of great importance, because the absence of uniformity can provide local variations in the thickness of the deposit in electrodeposition processes, uneven corrosion of the electrodes in industrial cells, lower energy efficiency of primary and secondary batteries, or a decrease in the spatial time yield of three-dimensional electrodes. In addition, an uneven current distribution may decrease the current efficiency of electrolytic processes (BORIN, 1986).

The main factors influencing the current distribution are:

- Cell geometry;

- Electrolyte and electrode conductivity;
- Activation overvoltage in the electrodes, dependent on the kinetics of the electrode;
- Overconcentration, controlled mainly by mass transport.

Depending on which of the factors is considered, the current distribution types can be classified into three modes as shown in Table 2.1.

The classification of current distribution into primary, secondary and tertiary results from the application of the fundamental laws of electrochemistry to technological systems.

Table 2.1 - Current distribution in three modes

Types of Current Distribution	Factors considered	Parameters involved
Primary	Absence of overvoltages.	Geometry (cell, electrode).
Secondary	Activation over-voltage with no concentration variations near the electrode.	Geometry, activation overvoltage, electrolyte conductivity and electrodes.
Tertiary	Overactivation and concentration.	Geometry, conductivity, activation overvoltage and concentration.

The current distribution will be uniform only if the entire electrode arrangement is geometrically similar. Figure 2.2 illustrates four electrode geometries. In the cell with flat and parallel electrodes (a), and with cylindrical electrodes (b), the current distribution will be uniform due to the symmetry of the arrangement. On the other hand, in cells (c) and (d) with different arrangements, the current distribution will not be uniform.

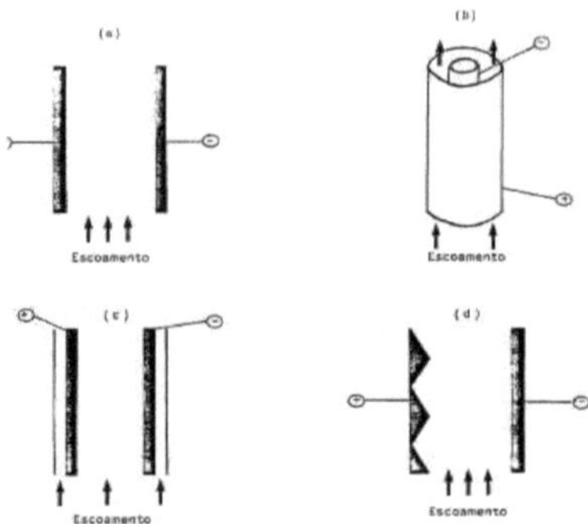

Figure 2.2 - Electrode geometry applied to electrochemical processes (BORIN, 1986).

In cell (c) the faces of the electrodes facing the cell walls are not isolated, allowing a small current flow through these faces (which in fact decreases the current efficiency for the process); in cell (d), for the most distant cathode points from the anode, the current flow will be lower than at the points closest to the anode (also decreasing the current efficiency).

One way used to increase the active electrode area is to place several electrodes side by side so that both sides are used. Another solution is to use reactors composed of concentric cylindrical electrodes. The advantage of this electrode arrangement, as well as the parallel flat electrodes, is that the current distribution is uniform due to the symmetry present. Therefore, the entire electrode area is active. Another alternative is to use porous electrodes, also known as volumetric electrodes, which began to be effectively applied in electrochemistry only after 1970.

The behavior of current density and electric field in an electrochemical cell with two electrodes was simulated by Pretorius (2015), using the COMSOL modeling program and is presented in Figure 2.3.

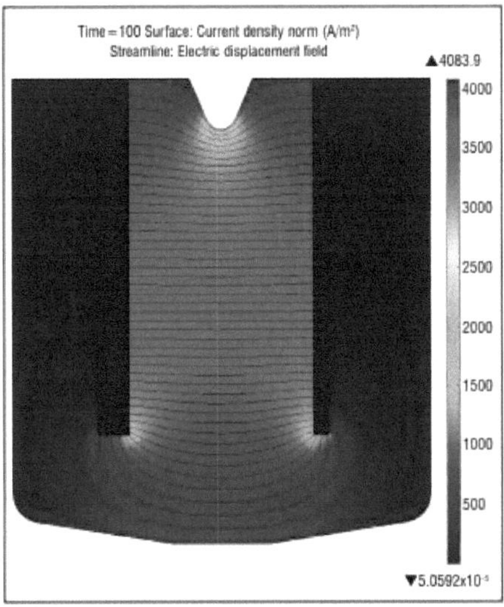

Figure 2.3 - Current density distribution and the electric field lines within the cell. Pretorius, 2015.

Current density values are high at the sharp edges of the electrodes due to the small surface area available. High current density values are also visible along the reactor because charged ions flow around this point to transit between the electrodes (PRETORIUS, 2015).

2.6.3. Faraday's Law

Michael Faraday (1791-1867) formulated two laws governing the quantitative part of the phenomena linked to electrolysis. Assuming that "the passage of an electric current from a metallic conductor to an electrolytic conductor, or vice versa, is always accompanied by an electrochemical reaction", Faraday proposed: "The magnitude of the chemical effect, in chemical equivalents, is the same on both the metallic surface and the electrolytic solution and is determined only by the amount of electricity that passes". That is, if an electron equivalent flows through an interface, a gram equivalent of the species involved in the electrolytic process should be oxidized or reduced. Where an electron equivalent corresponds mathematically to:

$$F = N. \text{ and } \rightarrow F = 96500C.mol\text{-}1 \qquad \text{Equation (12)}$$

Where F is the Faraday constant, N = Avogadro number, and is the charge of the electron. Therefore, in an electrolysis, the mass of material involved in each of the electrodic processes is given by the expression:

$$ma = \frac{Q.Ma}{n.F} \qquad \text{Equation (13)}$$

Where:

ma = mass obtained from the element (g)

Q = Load (C)

M = molar mass of element a (g/gmol)

F = Faraday constant = 96500C/mol of electrons

n = number of electrons involved in the reaction

This expression allows, for example, the determination of the mass and thickness of the deposited material assuming that the current density (current/area) is constant and known. Two types of processes can occur in the electrodes, the faraday and the non faraday. One type is that in which the loads are transferred through the metal/solution interface (electrode/electrolyte). This transfer of electrons causes oxidation or reduction, since these relations are governed by Faraday's law (the quantities of substances released or accumulated in the electrodes of a cell are directly proportional to the quantity of charge that passes through the solution), they are called faraday processes (BARD & FAULKNER, 2001).

2.6.4. Applications of Electrochemistry in metal removal

To design an electrochemical deposition process for precious metal recovery, it is necessary to have a model to describe the metal recovery (or the residual concentration of metal in the solution) as a function of operational parameters such as time (CHEN et al., 2005).

Electrochemistry deals with the transfer of charge at the interface between an electrically conductive material and an ionic conductor, as well as with the reactions within the electrolytes and the resulting balance. Cathodic removal of heavy metals shows several benefits in terms of cost, safety and versatility. The equipment required is only an electroplating effluent, an insoluble anode and a suitable cathode.

The recovery of heavy metals from aqueous solutions is of utmost importance in various industrial and environmental frameworks, from mining to wastewater treatment to waste recycling. Over the past two decades, extensive studies have been conducted on the recycling of electronic waste.

Several electrochemical approaches have already been studied for metal removal and recovery. Kongsricharoern and Polprasert (1996) used an electrochemical precipitation process to treat a residual water from galvanoplasty containing Cr concentrations of 570-2100 mg l-1. It was found that the Cr removal efficiencies were higher than 99% and the Cr concentrations in the treated effluent were lower than 0.5 mg/L. Kusakabe et al. (1986) studied the simultaneous electrochemical removal of copper and organic waste using an electrochemical cell. The final concentration of copper in the effluent was as good as 3 mg/L. A study by Hwang et al. (1987) showed that under strong alkaline conditions (pH $\geqslant$ 12), cuprocyanide ions could be oxidized directly and copper oxide precipitated at the anode.

Using the electrochemical approach to recover metallic ions in wastewater in its metallic state can be considered a relatively simple and clean method. Besides recovering metals in their metallic form, electrochemical treatment has several advantages: no extra chemical reagents are needed; no sludge production; high selectivity; low operational cost and possible decontamination of wastewater (LI et al., 2004; CHEN et al., 2005).

Even with so many advantages, there are some disadvantages among them: the deposition rate and the solution composition in some cases can cause the production of dendrites and loose or spongy deposits in the cathode and the interference of the hydrogen evolution reaction.

The recycling of obsolete Printed Circuit Boards is, at present, a relatively new activity in Brazil, although there are opportunities for expansion in this area. For example, gold, silver, tin and copper, among other metals, can be recovered through hydrometallurgical treatment of these electronic waste, followed by electrochemical processes (VEIT et al., 2005, 2006; MANETTI et al., 1995, 1996).

In view of all the reviews presented, the study of the recycling of metals from these wastes has reached the point of being completed. Studies already carried out show that there is no efficiency in treating all metals in a single process, or parameters. What actually occurs is the removal of one, two or three different metals for each studied technique, or for each modification of parameters, such as time and current applied. This study is based on techniques already addressed, with innovations, however, obtaining different results from those shown in the literature.

In this study, we will address the removal of copper, silver and tin, as well as other metals, focusing on the recycling of obsolete computer Printed Circuit Boards from the Federal University of Rio Grande do Norte - UFRN. There are countless works performed to remove copper from this type of material, a considerable amount of silver removal and a very small number dealing with the removal of tin. However, some techniques are capable of recovering high concentrations of metals, but not to admissible levels or including more processes, thus increasing the costs of metal recovery. Electrochemical techniques are an excellent alternative for metal recovery due to their low operating cost and the possibility to handle low concentration electrolytic media, which allows to reduce their environmental impact.

Some authors have performed the treatment of this material for extraction and removal of these metals in solution, by methods similar to those used in this thesis. Dutra, 2014 - presented a new process for the production of copper from Printed Circuit Board (PCB) concentrated powder. The results indicated that the agitation and temperature increase favoured the recovery of copper. However, agitation proved to be the most important parameter to increase the efficiency of copper recovery from PCBs. At 25°C, an apparent 96% copper recovery was achieved after 15h of electrolysis at an agitation speed of 415 rpm, while at 40°C and without agitation the recovery was only 90%. Martins, 2009 - performed the Leaching of Printed Circuit Boards (PCBs) from obsolete computers for extraction and recovery of tin and copper through leaching followed by precipitation. The leaching that showed the best results was using 3.0N HCl + 1.0N HNO3, with an extraction of 98% for Sn and 93% for Cu. Precipitated were obtained at different pH values, neutralizing the leachate with NaOH. The leaching system 3.0N HCl + 1.0N HNO3, presented the highest values of recovery of dust (84.1% for Sn and 31.9% for Cu), as well as leaching (85.8% for Sn and 34.3% for Cu). Park and Fray, 2009 - recovered precious metals from Printed Circuit Boards. Regia water was used as a leachate and the relation between metals and leachate was fixed at 1/20 (g / mL). Silver is relatively stable, so that about 98 % by weight of the input was recovered without further treatment. Palladium formed a red precipitate during dissolution, and the precipitate amount was 93 % by weight. A liquid-liquid extraction with toluene was used to extract the gold selectively. In addition, dodecanethiol and sodium borohydride solution were added to form gold nanoparticles. The gold extracted was about 97 % by weight. Lee (2003), studied the recovery of copper, tin and lead from the nitric solutions used in the recording of the Printed Circuit Board and the regeneration of this solution. Solvent extraction, pickling, electrodeposition, precipitation and cementation were performed. The pure nitric acid solution was removed by distilled water. After the extraction of nitric acid, pure copper metal was obtained by electrolysis and tin ions were precipitated adjusting the pH of the solution with Pb (OH). Lead metal with 99% purity was obtained by cementation with iron powder.

Veit et al. (2006) used electrodeposition to recover the copper present in printed circuit boards. The authors used mechanical processing (comminution, particle size, magnetic and electrostatic separations) to concentrate most of the metal in a single fraction, which after leaching, was electrodeposited, obtaining a 98% recovery percentage in most cases.

Xiu and Zhang (2009) studied the Cu deposit from leached solutions of printed circuit boards (PCBs) using supercritical water as an oxidizing agent. In the electrochemical part (deposition), they obtained recovery of 84.2% of the Cu deposit with current density of 20mA / cm2, during 11 h of treatment in HCl 1M, which was used as electrolytic medium. During the electrolysis process, it was observed that Cu deposits were partially oxidized in Cu2O, which represented an interfacial limitation in the oxidation or deposition processes.

Regarding copper recovery, it is undoubtedly a more explored and studied process than the recovery of other metals, the existing bibliography already has practically all the existing alternatives covered, and the existing industrial applications are already quite optimized, giving no margin for the emergence of new processes (GOUVEIA, 2014).

For Gouveia (2014) it is interesting to approach the recovery of other metals, such as the study of platinum, silver and tin group metals, since they also have high market value and can be found, in different quantities, which are not negligible, in Printed Circuit Boards.

Some recent studies conducted by Lekka et al., (2015) reported the feasibility of Au recovery through deposits of a solution obtained from leaching PCI's in aqua regia. Synthetic solutions were also used, intentionally adding specific concentrations of different metals to observe the influence of Cu on system behaviour. It was found that Cu caused a shift in the Au reduction potential to more negative potentials, as well as resulted in a significant increase in current due to the high Cu concentration. The latter, together with the low Au concentration present in the solution, caused a situation where the peak gold reduction could not be distinguished in voltamograms. For this reason, the solution was enriched with Au. Using this approach, a reduction potential for this metal was found to be 0.55V vs Ag / AgCl / KCl3M. Thus, it was established that the concentration of metals in the system will cause significant changes in the reduction potential of precious metals, making it difficult to recover them selectively.

Fogarasi et al., (2015), studied Cu deposit from a solution obtained from PCI's chemical (acid) leaching. They proposed a chemical and electrochemical process in a reactor using HCl and FeCl3 as electrolytes, with the objective of dissolving the plastic parts of PCI. A 99.95% purity Cu deposit with a flow rate of 400 mL / min and a cathode potential of - 0.135 V vs. Ag / AgCl / KClSAT was achieved. A solid residue was obtained in sludge containing important metals such as Ag, Au, Sn, Pb, Zn, Fe and Ni, which will require further steps conducted for its recovery. Fogarasi also pointed out that there are still several problems to be solved in terms of metal recovery from electronic waste, such as improving selectivity and reducing reagent consumption.

Regarding the extraction of precious metals from residual solutions containing metals by means of electrochemical routes, several authors have discussed data difficult to interpret from traditional analyses (mainly cyclic voltammetry) due to the presence of dissolved species of metals that present multiple states of oxidation, which prevents the true reduction mechanisms of each other. A common solution to this problem is the use of synthetic solutions which, although they simplify the analyses, do not always accurately describe the behaviour of real waste solutions. Some authors have proposed deposition experiments using real waste solutions under the best conditions found for synthetic solutions. These experiments prove that the use of synthetic solutions is somehow misleading.

By Cyclic Voltametry, the results of deposition of this type of waste result in an energy competition due to galvanic interactions, which prevents these noble metals from being deposited anymore, as verified in the studies conducted by Cortés-López et al., 2017. Another study carried out by Reyes-Cruz (2002) shows that Cu interference cannot be completely avoided. He observed that for potentials above -0.25V, it means an increase in cathodic currents as the cathodic potentials imposed increase. This behavior suggests a deposit of the species with the highest concentration in the solution, such as Cu. Potentials of -0.5V to -0.9V, the increase in the reduction current is very high, indicating that there is enough energy to reach a high Cu deposit, besides the evolution reaction of hydrogen. Therefore, in the present study it was necessary to characterize deposits obtained in order to define which species are preferably deposited in each potential.

We verified by this study with Cyclic Voltametry, that even considering that determining element is deposited in a certain potential, in this type of residue, there is a great possibility of more elements being together in this same deposit, which leads the great majority of the authors to use synthetic solutions to avoid interference from other elements. In our study, Cyclic Voltametry was not performed precisely because of the difficulty of interpreting them, however, the deposition potentials were observed during the reactions.

Chapter III

Methodology

3. Methodology

For a better visualization of the development of this research, the experimental process has been divided into steps and is shown in the flowchart of Figure 3.1.

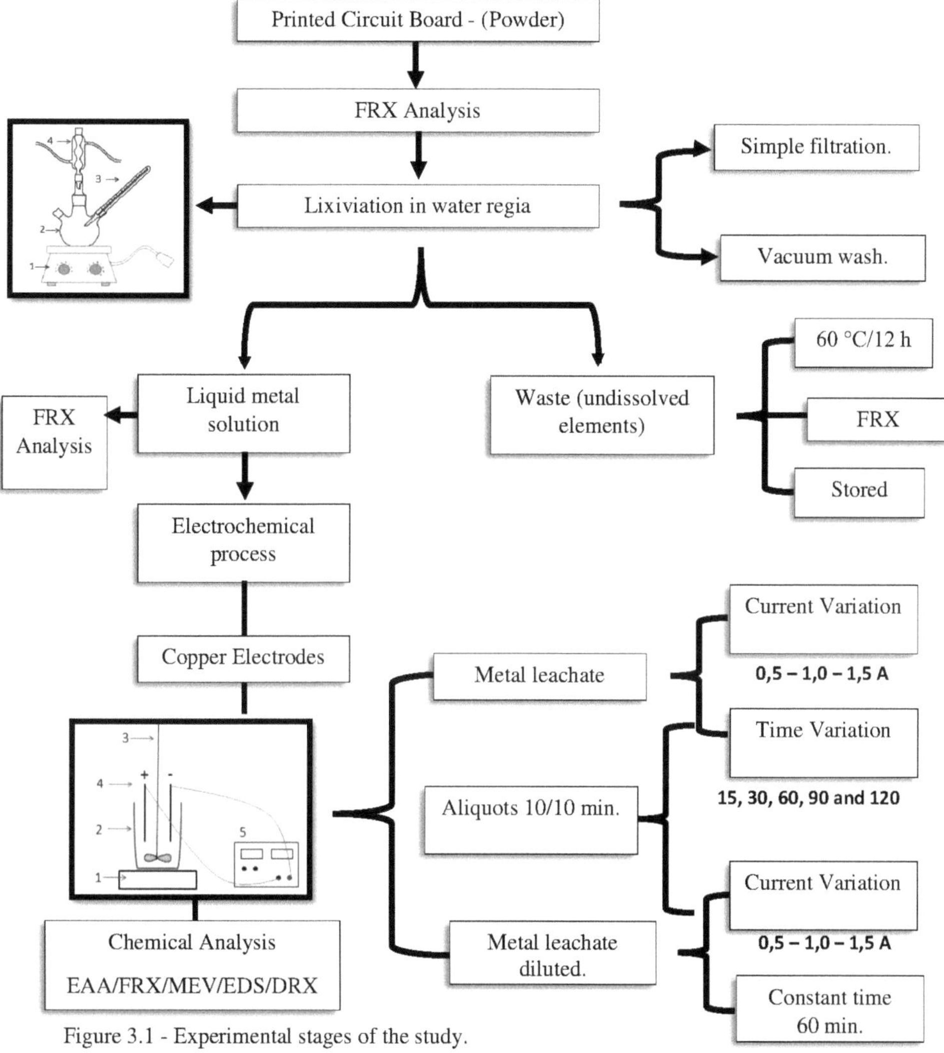

Figure 3.1 - Experimental stages of the study.

3.1 First step: dissolution of ICP dust by leaching with aqua regia.

3.1.1. printed circuit boards (PCBs)

The Printed Circuit Boards used in this study are desktop computers and were part of the Department of Materials and Heritage of the Federal University of Rio Grande do Norte - UFRN, not being specified their brands and models, and bearing in mind that they are some very old. In this study, only the substrate of the boards was used, but it will always be referred to here as Printed Circuit Board (PCB). Figure 3.2 shows a sample of the powder obtained from the Printed Circuit Boards used in this study, previously passed through mechanical processing.

Figure 3.2 - Dust sample obtained from Printed Circuit Boards.

The objective of the fragmentation of these residues carried out by Medeiros, 2015; Melo, 2017, was to reduce the size of the material in order to facilitate chemical processing since the size of the particles has an influence on the extraction speed of the metals present in this residue.

Kirchner studies (1999) have shown that fragmentation should allow for particles of less than 1 mm in diameter (typically 0,5 mm or less, depending on subsequent separating technologies) to achieve an efficient material release. Mechanically bonded plate components can be separated and released by efficient grinding. However, bonded materials in more cohesive forms, such as composites or very adherent coatings, will not be separated by this type of treatment, nor will separation of chemically bonded elements (such as metals in alloy form) be possible by physical processing (OGUNNIYI, 2009).

The granulometry of the material, as well as the leaching temperature and time, among other parameters, have an influence on the leaching performance of PCI's, being necessary for its optimization to maximize the recovery of metals in attractive technical-economic conditions. For research purposes, approximately 2.07 Kg of dust from these residues was used. The granulometric fraction of the sample obtained by the authors and used in this study is ≤ 1mm.

The procedure took place as follows: the powder obtained from the plates, composed of several metals of different concentrations, was digested in aqua regia and then characterized by X-ray Fluorescence Spectroscopy and used in the electrochemical process.

3.1.2 Characterisation of printed circuit boards (PCBs)

The PCI's characterizations were performed in order to obtain the actual quantity of each element in the sample. In the digestion of PCI's a solution of aqua regia was used, composed of nitric acid and hydrochloric acid (25% HNO3/75% HCl) in a ratio of 1:3, based on previous studies performed by Veit et al., 2005, Petter, 2012, among others and experimentally proven in this work. This solution demonstrated a high efficiency in the solubilization of several metals. In this step, a closed system with reflux was used, as shown in Figure 3.3, to avoid losses of reagents to the environment. Conditions such as time and ambient temperature were analyzed to verify the best leaching system, the results are presented in the results and discussion section.

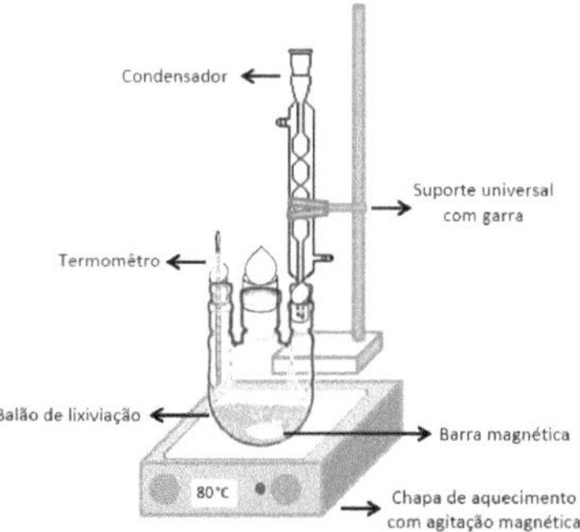

Figure 3.3 - Chemical digestion of water regia printed circuit boards in a reflux balloon - Silvas, 2014.

In the process of dissolution by leaching a volume of 500ml of royal water and a solid sample of approximately 25 g of PCI's ground for each assay performed with constant magnetic stirring were used. At the end of each leaching, the system was filtered, via simple filtration with quantitative filter paper, the leached extract and the paper residue were forwarded for chemical analysis by X-Ray Fluorescence Spectroscopy (XRF).

The solution, which represents the soluble metals, was sent to chemical analysis by X-Ray Fluorescence Spectroscopy (FRX), where the research of the metals present in solution was performed, and the solid residue retained in the filter paper, insoluble portion, represents the fraction of polymers and ceramics, was dried in an oven at 60°C for 24 hours, then weighed and sent to chemical analysis by the same technique mentioned above. Figure 3.4 provides a summary of the leaching step.

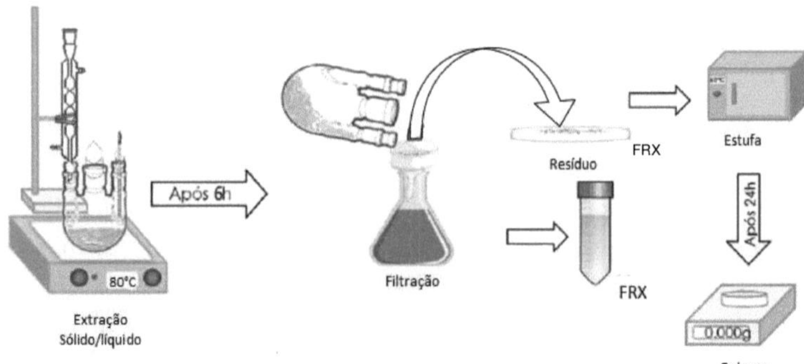

Figure 3.4 - Stages of leaching. Adapted from Silvas, 2014.

In order for the leaching to reach maximum metal dissolution, some parameters such as time and temperature had to be optimized. The conditions investigated are presented in Table 3.1 and were based on studies already performed by other authors.

Table 3.1 - Parameters studied in leaching.

Test	T (°C)	t (h)
1	25	6
2	25	24
3	80	6
4	60	6

The fraction of leached mass (L) was obtained by gravimetric analysis and calculated by Equation (14).

$$L = 100x\frac{mi - mf}{mi}$$

Equation (14)

Where;

L represents the leached metal mass (%)

mi represents the mass of PCI measured before the leaching test (g)

mf represents the mass of PCI measured after the leaching tests (g)

3.1. Second step: Production of copper electrodes and construction of the electrochemical cell for the metal deposition process.

3.1.2. Production and characterization of electrodes

The electrodes used in this process are made of copper and manufactured with motor winding wires, purchased in the commercial industry. They consist of a cylindrical shaped copper rod, some measuring 20 cm and others 25 cm long and 3.5 mm in diameter. Initially the electrodes, already with defined sizes, were sanded to remove the resin that surrounds the wires avoiding oxidation, and then some were manually molded to be straight, and others were molded so that one of the tips formed a circle, as shown in Figure 3.5.

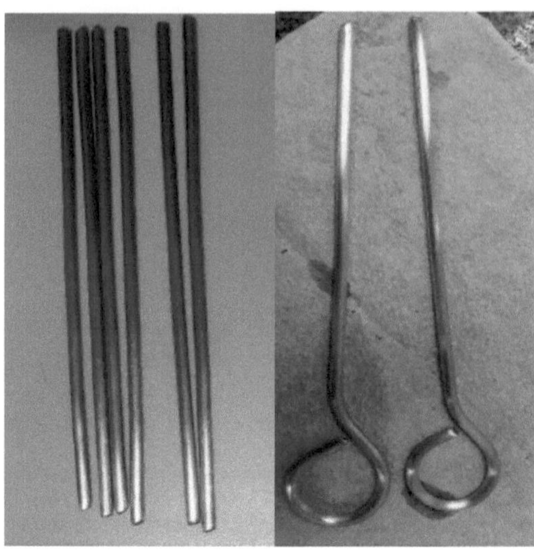

Figure 3.5 - Electrodes used in the electrodeposition process.

After this step, they were washed with sponge and detergent several times, rinsed with distilled water, dried with paper towels and stored in plastic film to avoid oxidation. The electrodes were characterized by X-ray Fluorescence Spectroscopy. Later they were used in the process of silver, copper and tin deposition, which will be described in the next topics.

3.1.2. Electrochemical cell

The cell used for the electrochemical process, was acquired in the commercial industry. The material of the cell is glass, with capacity for 1 liter of solution. The mechanical agitator used in the process with stirring, contains a motor with rod and Teflon blades to avoid oxidation in the solution. A Minipa Mps-3005 30V/5A Digital power supply was used to supply current to the system. Figure 3.6 presents the electrochemical cell used in this study for the electrodeposition process of metals.

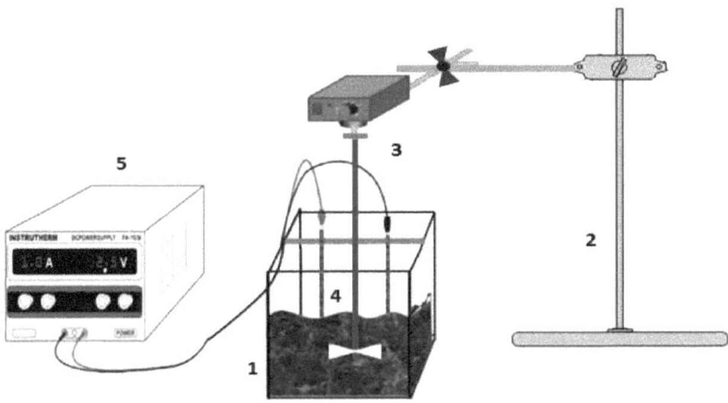

Figure 3.6 - Electrochemical cell. 1) electrochemical reactor, 2) support, 3) mechanical agitator, 4) copper electrodes, 5) power supply.

3.2. Third step: electrochemical process with the actual solution and optimization of parameters.

The electrodeposition of metals in this study was performed using the galvanostatic method (constant current), because the experimental module is simpler (cell of two electrodes) and often the electrodeposition is faster. For the deposition of metals, some parameters were explored to evaluate the process of removal and deposition of metals from a leached solution of Printed Circuit Boards.

In the first stage of metal extraction, starting from the solution leached with royal water, real solution, experimental studies were performed at current densities of 0.11 - 0.22 and 0.33 A / cm2 and constant time of 60 minutes. The current density corresponds to the ratio of the direct current applied by the total area of the active surface of the electrode. With the optimized current, the second step of these experiments was performed for different times 15, 30, 60, 90 and 120 minutes at a constant current density of 0.22 A / cm2. Electrodeposition tests were also performed for the diluted solution of 1:3 and 1:7, and all the current densities used previously were also tested, using a constant time of 60 minutes.

3.3.1 Electrodeposition testing

After the initial characterization of the samples, solubilization, filtration and chemical analysis, electrodeposition tests were performed in order to remove the dissolved metals from the leachate. The leached liquor from PCI's has a pH below zero. In this step, all tests were performed with the real solution, that is, without dilution or addition of other chemical reagents. Its conductivity was measured using a bench-top conductivimeter, of the Konduktometer SCHOTT brand, with a value of approximately 465 mS/cm, without the addition of any electrolyte.

The electrodeposition experiments were conducted in a batch reactor. A pair of copper electrodes, with an active cathode area of 8 cm2, was inserted vertically and connected to a DC power supply. The electrodes were separated by plastic spacers at a distance of 3 cm and then the effluent was added to the reactor and the system was maintained under constant stirring by means of a mechanical stirrer at a speed of 550 rpm for the stirring process. The volume used in each test was 400 ml. Before each test, the electrodes were washed, dried and weighed as described in the previous section. After each test, the electrodes were dried in an oven at 60 °C and then weighed to verify the loss and gain in mass after the end of each process.

To obtain the metals in the working electrode, electrodepositions using some deposition currents were performed. The knowledge of the optimal currents to be applied is not a trivial step, considering the complexity of the solution and the absence of references in comparable systems, the current to be used is a completely unknown variable, which was not possible to perform an experimental planning, for this reason, this step consisted of previous experiments, which thus made it possible to delimit a range of current that provided deposits in the electrode, this range of current could only be determined with the aid of the X-Ray Fluorescence (XFR) technique, which was readily available after each test experiment.

In this electrodeposition stage, with a pair of straight electrodes, being one cathode and one anode, optimization tests of the conditions to be applied were performed, first varying the current applied between (0.5 A; 1.0 A and 1.5 A). To study the effect of current density on the composition of the deposited material, the electrolysis time used in each density was 60 minutes. This time was randomly chosen. The current densities were 0.11 A/cm2; 0.22 A/cm2 and 0.33 A/cm2. Once the current that presented the most significant results was defined, it was necessary to optimize the time of metal deposition in the process.

The deposition time varied between 15, 30, 60, 90 and 120 minutes and the current density was kept constant at 0.22 A/cm2. All experiments were performed in a non-sequential way, i.e., after each test, the aliquots collected during the process, the powder deposited in the cathode and the cathode were sent for chemical analysis by X-Ray Fluorescence (XFR) analysis and another process was started with a new solution. In all FRX analyses, a scan of all the elements of the periodic table was performed.

Table 3.2 presents a summary of the trials performed at this stage. All trials were repeated and between two and three trials were performed for each experimental condition, where the results are an average of the values obtained. Aliquots were collected at time intervals of 10 minutes to follow the progress of the reproducibility of the experimental results by process time.

Table 3.2 - Summary of conditions applied in the 1st electrodeposition tests performed.

Current (A)	Agitation (RPM)	1st Time (min.)	2nd Time (min.)
0,5	With	60	15, 30, 60, 90, 120
	No		
1,0	With	60	15, 30, 60, 90, 120
	No		
1,5	With	60	15, 30, 60, 90, 120
	No		

Then, it was found that there was no good reproducibility of the results under experimental conditions when the current values of 0.5 and 1.5 A were applied, and the next tests were performed for the current of 1.0 A. Starting from the point of the already optimized current, the electrodeposition tests were performed in times of 15, 30, 60, 90 and 120 minutes, pH below zero, with agitation (550 rpm) and without agitation, with the objective of optimizing the best time of removal and analyzing the most favorable conditions for metal deposition.

3.2.2. Chemical Analysis

Chemical analyses were performed by X-ray fluorescence in 720 Shimadzu spectrometer with EDX (XRF) detector and Scanning Electron Microscopy (SEM) under Carl Zeiss EDS-coupled microscope. The analysis of silver concentration in the solution at the end of each reaction was measured by Atomic Absorption Spectroscopy (EAA) on an

AA240 Varian spectrometer. The analytical curve for the determination of Ag covers the range from 0 to 15 mg L-1 in the air-acetylene flame, with the following points 0; 3; 6; 9; 12; 15 mg L-1. There was linearity between the data with r = 1. X-ray diffractometers were obtained using a Bruker D8 Advance Davinci automatic diffractometer with a continuous, step-by-step scanning device. Diffraction patterns were measured in a range of 0-80° 2θ using Cu K radiation of 40 kV and 40 mA with a scanning rate of 5° per minute.

3.3. Fourth step: electrochemical process performed with the solution diluted in varying proportions

3.4.1. diluted solutions

Following the same methodology, the leached liquor solution was diluted in a proportion of 1:3 and 1:7, being 1 volume of aqua regia and 3 volumes of distilled water (1:3) and 1 volume of aqua regia and 7 volumes of distilled water totaling a volume of 500 ml to investigate the extraction of metals in the presence of water, simulating a real contaminated effluent of these metals. This volume was added to the reactor and the system was maintained without agitation and with agitation for 60 minutes at 550 rpm. During the reaction, aliquots were collected with a plastic syringe every 10 minutes to follow the evolution of metal removal. The pH of the solution still presented values below zero.

The same arrangement in the cell, with two copper electrodes, was maintained, and the current densities were the same as already described. In these tests, neither longer nor shorter times than 60 minutes were analyzed, however, the reactions were performed with agitation and without agitation.

3.3.2. Chemical Analysis

The collected aliquots were analyzed by X-ray Fluorescence Spectroscopy in 720 Shimadzu spectrometer with EDX detector (XRF) and Atomic Absorption Spectroscopy (EAA) in AA240 Varian spectrometer. The cathode deposits were analyzed by Field Emission Scanning Electronic Microscopy (SEM-FEG) - with Dispersive Energy - EDS.

Tin determination, at the end of each process, without mechanical agitation, was performed by atomic absorption spectrophotometry at 224.6 nm of wavelength. A standard gas-acetylene flame burner with gas flow of 4 L/min and tin lamp current of 7mA was used.

X-ray diffractometers were obtained using a Bruker D8 Advance Davinci automatic diffractometer with a continuous, step-by-step scanning device. Diffraction standards were measured in a range of 0-80° 2θ using Cu K radiation of 40 kV and 40 mA with a scanning rate of 5° per minute.

3.4. Fifth step: increase and variation of the active area of copper electrodes

3.5.1. increase in active electrode area in different arrangements

Tests were performed increasing the active area of the electrode and consequently an increase in current density in the reactor. The conditions used were the same as those presented in Table 3.2, for current and time, the other parameters, as time remained constant. The electrode arrangements were modified to verify the influence of the arrangement on metal extraction, especially silver. In the first arrangement, two electrodes with circular active area were used, being one anode and one cathode, and in the second, a parallel electrode arrangement with straight active area, being two anodes and two cathodes, both connected to a direct current source. Borin (1986), reports several electrode arrangements for metal removal in effluents. Among them are the parallel arrangements.

In this research, electrochemical processes were performed using the parallel electrode arrangement, which works as a way to increase the active cathodic area. In this arrangement, 4 electrodes were placed in the electrochemical cell side by side, so that both pairs were used. A distance of approximately 3 cm was maintained between them. This bipolar arrangement was electrically connected to the power supply so that the same current was flowing through all the electrodes. This type of arrangement played an important role in this study due to the fact that it is ideal for processes in which both cathodic and anodic reactions occur in the same type of material.

3.5. Electrode weight

To quantify the gain and loss in mass of the electrodes, they were weighed three times before and after each test. The calculation was performed by the difference of initial and final weight and the results show an average of these weights.

In all reactions, at the end of the processes, with and without agitation, the electrodes are removed and taken to dry in the oven at 60°C for half an hour. Afterwards, the electrodes are removed from the oven and weighed to check the mass gain. The tank is easily removed by swinging the part with the tank inside an ependorf and forwarded to chemical analysis by FRX and MEV-FEG - EDS. When starting another process, new electrodes are used.

The gravimetric method was used to calculate the actual concentration of silver and tin present in the cathode. Using Faraday's law, it was then possible to obtain the respective masses according to the deposits. Faraday's law predicts the mass of a species that can be deposited by the amount of electricity that passes through the solution and is expressed by Equation (15). Knowing that the amount of charge that passes through the cell is obtained through the ratio shown in Equation (16).

$$ma = \frac{Q.Ma}{n.F} \qquad\qquad \text{Equation (15)}$$

$$Q = j.A.t \qquad\qquad \text{Equation (16)}$$

Being;

Q = load (C);

A = area (cm2);

j = current density (A/cm2);

t = time (s).

Chapter IV

Results and discussions

4. Results and discussions

4.1 Processing of the powder by leaching with aqua regia

4.1.1 Dust characterisation

In the composition of the powder, 17 elements were identified. It can be clearly seen from Table 4.1 that, except for hazardous substances, there are many valuable materials contained in PCIs, which are worth recycling as reported by Zhou and Qiu, (2010) in their studies. Table 4.1 shows the mass percentage of the main elements that make up the Printed Circuit Board used in this study.

Table 4.1 - Composition of Printed Circuit Boards.

Element	%	Element	%
Cu	24,403	Pb	0,536
Sn	20,178	Fe	1,483
Br	18,264	Ag	0,635
Si	16,919	Cr	0,329
Ca	13,485	Ni	0,151
Al	3,404	Ti	0,147
Ba	1,963	Mn	0,032

Based on the percentage of mass, copper was found to be the dominant metal species (24.403% by weight). In addition to copper, FRX was also able to detect the presence of some valuable metals in the residues including silver (Ag), nickel (Ni) and titanium (Ti) as observed by Zhou and Qiu (2010).

It should be kept in mind that all electrical and electronic waste are mixtures of different equipment and components, which generally have a random chemical composition depending on their source (PC, mobile phone, music players, etc.).

This material is considered homogeneous due to the quartedness of the sample.

4.1.2. Leaching of metals in acid medium (HCl and HNO3)

In table 4.2 the leaching yields for most metals are shown. Test 1 shows leaching at room temperature for 6 hours, test 2 at room temperature for 24 hours, test 3 at 80°C for 6 hours and test 4 at 60°C for 6 hours, all tests were performed with aqua regia.

Table 4.2 - Leaching yield of metals for the tests performed.

Test	L (%)
1	23,86
2	26,27
3	35,01
4	32,17

With the water regia leaching, and due to the presence of nitric acid and its oxidizing effect, the behavior observed in the 4 tests analyzed is distinguished according to the variables used, and there was a better leaching performance for most metals in test 4. It was also found that the leaching performance in general is favored by the increase in temperature totaling 35.01% of leached mass when the system was at 80°C for 6 hours.

Thus, the leaching yield obtained for copper in water regia tests was 100% at 80ºC, and at all times and temperatures performed, its efficiency was over 90%, being therefore a very effective leach for this metal.

For most metals, this leachate has proved to be very efficient, as for iron, tin and lead with yields above 70 %. Titanium, calcium, chromium, zinc and nickel had similar yields to copper, above 95%.

As for silver, a good yield was also obtained, above 70% for a leaching at 80°C. Even obtaining high metal extractions, metal fractions were still detected in the residues, as shown in Table 4.3.

Table 4.3 - Residual composition of the sample on filter paper after the leaching process.

Elements	Test 1 %	Test 2 %	Test 3 %	Test 4 %
Br	30,00	31,80	51,00	52,00
Si	28,00	42,43	38,00	34,00
Ca	24,00	7,87	0,06	2,00
Al	14,00	13,94	7,00	8,00
Ba	2,00	1,21	0,32	1,00
Fe	1,10	1,07	0,44	1,00
Sn	0,51	0,69	3,00	2,00

Ti	0,33	0,43	0,20	0,20
Mr	0,25	0,23	0,02	0,06
Pb	0,12	0,11	0,19	0,19
Cu	0,09	0,08	0,07	0,07
Ag	0,05	0,07	0,15	0,27
Cr	0,03	0,03	0,00	0,01
Zn	0,01	0,01	0,01	0,01
Ni	0,00	0,01	0,00	0,01

The percentages of metals presented in table 4.4 refer to the leached system at a temperature of 80°C and 6 hours of reaction, this being the solution used in the electrochemical process.

Table 4.4 - Main elements present in solution.

Elements	%	Elements	%
Cu	40,711	**Fe**	1,195
Cl	24,904	**Ag**	1,539
Sn	17,265	**Al**	0,601
Ca	7,326	**Ni**	0,209
At	3,638	**Cr**	0,179
Pb	1,966	**Ti**	0,154

The results shown in Table 4.4 indicate quantities of copper and tin much higher than the other elements. Due to its high value, the presence, although small, of precious metals such as silver makes the recycling of this waste economically very interesting.

4.2. Electrochemical process with the actual solution and optimization of parameters

4.2.1. Effect of current density on silver removal and deposition time

In all electrochemical processes, current density is a very important parameter to control the reaction rate within the electrochemical reactor. The current density, in this case, determines the production rate of $Cu2^+$ ions released by the anode. Figure 4.1 shows the electrodeposition of the metals after 60 minutes of reaction at densities of 0.11 A/cm2; 0.22 A/cm2and 0.33 A/cm2, and constant time of 60 minutes.

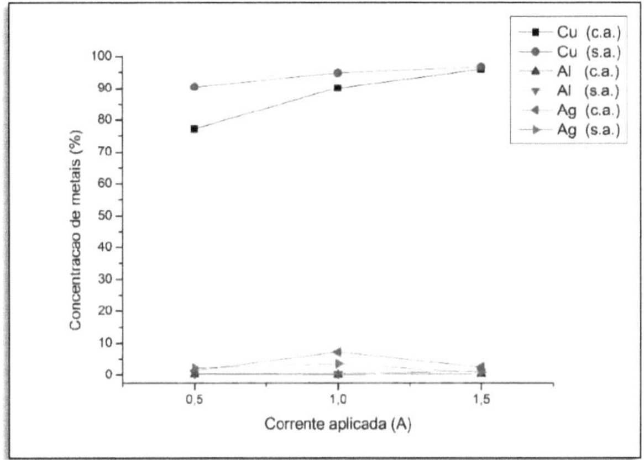

Figure 4.1 - Electrodeposition of metals after 60 minutes of deposition with and without mechanical agitation.

It is possible to observe in Figure 4.1 that the concentration of metals, mainly of silver deposited is related to the current applied, but does not present a linear behavior. The increase of copper concentration in the solution, with the increase of electric current and rapid dissolution of the anode, leads to an increase in the number of adsorbed ions at the electrode interface and leads to an increase of electric polarization. The higher the polarization, the thinner the layer deposited in the cathode, however, the smaller the removal, verified this behavior in Figure 4.1.

In 60 minutes of the deposition process, it was found that the increase in current density leads to a lower current efficiency, thus affecting the amount of metal deposit in the cathode. An increase in the current density leads to an increase in the excess of cathode energy, increasing the activation of the reactions on the electrode surface, which in turn causes an increase in the content of some species and a decrease in the content of others, which in fact is verified in the present study.

The electrodeposition process with continuous mechanical agitation shows better deposition rates, especially of copper and silver, at the three current densities. A satisfactory result was presented when the process operated at a density of 0.22 A/cm2 and 60 minutes, reaching a concentration by weight of 7, 15% of silver deposition, for the other current densities 0, 11 and 0.33A/cm2, 1.5% and 2.5% respectively were obtained. For the electrochemical reaction without continuous agitation, it was observed that in the silver deposition when the applied current was 1.5 A, its mass was less than 1%. For currents of 0.5 and 1.0 A, both without stirring, a silver deposition of 2.3% and 3.6% respectively was obtained. In all variables studied, copper deposits were above 90%, except for the 0.5 A current and without agitation. In these parameters operated in the electrochemical cell, the main reaction that occurred in the anode was to the dissolution of copper as Cu2+, and in the cathode occurred to the reduction of silver together with other metals; what in fact happens with the noble metals, where copper in the electrochemical reactivity series is capable of reducing silver in aqueous solution. In view of the results obtained with the variation of the current density, it was fixed for the next experiments the one that presented the best results in relation to the deposition of silver, which was the current density of 0.22 A/cm2.

From this point on, the deposition time was varied to investigate the influence of this parameter, since it is directly related to electric energy consumption. An observation to be made before these results, is that in fact there is the deposition of a layer of metals in the cathode, however, observing visibly during the process, a large amount of this deposit falls in the solution, and dissolves again. This low adherence to the cathode, mainly in the reactions without agitation, is being directly influenced by hydrogen production.

Figure 4.2 shows deposition rates of metals for both processes, with and without mechanical agitation, at different times of 15, 30, 60, 90 and 120 minutes and constant current density equal to 0.22 A/cm2.

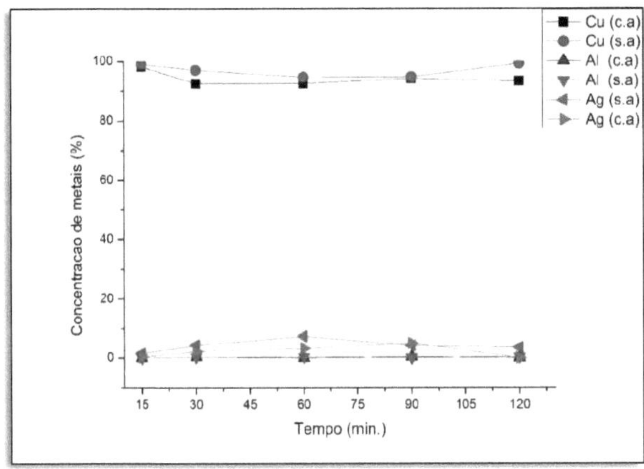

Figure 4.2 - Electrodeposition of metals for different times, with agitation and without mechanical agitation.

Analyzing in Figure 4.2, it was observed that the process without agitation presents deposition rates of aluminum, copper and silver, which vary in the times of 15, 30, 60, 90 and 120 minutes, presenting concentrations of 0,28%, 2,2%, 3,6%, 5,0% and 0%, respectively for silver, and in all times, the concentrations obtained of copper were above 90% and for aluminum below 1%. The value of 0% in the time of 120 minutes in the silver concentration occurred because the anode, after 100 minutes of the process, was completely oxidized, at this moment there was no longer a balance in the transference of loads in the solution, since the anode was no longer in contact with the solution making it impossible the passage of the current and all the cathode deposit fell in solution and dissolved again.

In the process with agitation, the deposition of copper and silver was shown in a significant way, presenting deposition rates above 1% in all analyzed times, being 1, 4% - 5,0% - 7,2% - 4,2% and 3,3% respectively for silver, and in studied times copper presents concentrations above 90%, as well as aluminum below 1%, as shown for the process without agitation. The times of 15 and 120 minutes presented different behaviors in relation to the deposition of metals, and in the process without agitation there were unexpected behaviors in 120 minutes, damaging the deposition of silver. In both processes the silver deposition rate in 15 minutes was low, being insignificant in the process without mechanical stirring, however, for electrodeposition in 120 minutes time and with stirring removal rates above 3% were achieved.

During the process, it was also observed that part of the metals adhered to the surface of the electrodes fall into the solution at certain times, dissolving again, making it difficult to collect them at the end of the process. This occurs because the solution is extremely acidic. Another possible hypothesis for the low removal of silver and other metals is the small active area of the electrodes used. Silver as well as copper are being deposited in the form of chlorides, silver chloride and copper chloride, as well as copper and metallic silver, verified in the analysis of X-Ray Diffractometer - (XRD), shown in Figure 4.3. Figure 4.3 presents the XRD of the cathode deposit, in a 30-minute reaction with and without mechanical agitation.

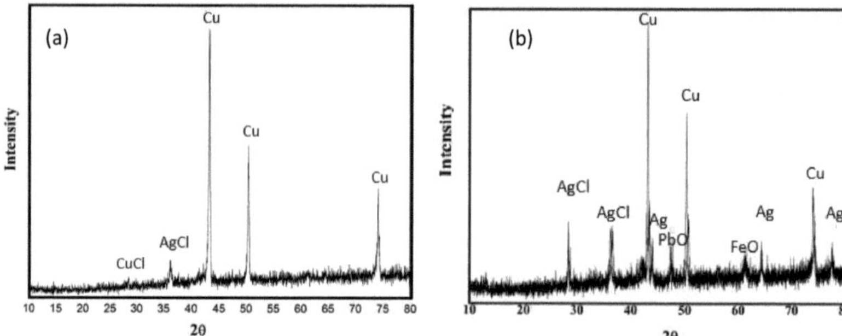

Figure 4.3 - X-ray diffractogram of the deposit formed in the cathode, (a) with agitation and (b) without agitation in 30 minutes of reaction.

One parameter that deserves to be highlighted in these analyses is the continuous mechanical agitation. The agitation of the solution has a significant effect for both cases studied, favoring the deposition of metals in the cathode. The agitation rate leads to an increase in the transfer rate of the ions present in the electrolyte and, therefore, their diffusion is facilitated from the solution to the cathode surface, which can also be attributed to the mass transfer in the solution, observed also by Torabinejad, et al., (2016) in their studies. The significant decrease in current efficiency, observed mainly when applying the 1.5 A current, can be observed with the decrease in the rate of metal deposition on the cathode surface, which in fact was verified. In this case, the evolution of hydrogen is the dominant reaction in the cathode, hindering the deposition process as well as the diffusion of ion species in solution. It is evident in both processes that the applied current was effective in all reaction times, which is the large objective of this variable time in order to minimize as much as possible the energy consumption in the process. The time between 30 and 60 minutes for this process would be sufficient for considerable removals of metals, including silver in this solution.

Even though the valuable removal of silver has already been identified, there are other metals that perhaps also have an affinity for the electrode used and have been deposited together with it, among them copper as already shown, other metals such as zinc, aluminum, lead, iron and calcium have presented small deposits at specific times, however they were well below 1 and are not shown. Table 4.5 shows the concentration of each element at all process times, being 15, 30, 60, 90 and 120 minutes operating with constant current. As already mentioned, each time studied presents a variation in deposition, with copper and silver being the metals of greatest occurrence in all times studied.

Table 4.5 - Concentration of metals present in the cathode.

Metals	Time (minutes)					
	0	15	30	60	90	120
	%	c. a. /s. a.	c. a. /s. a.	c. a. /s. a.	c. a. /s. a.	c. a. /s. a.
Cu	40,71	98,28/99,12	92,64/97,10	92,65/94,74	94,31/94,81	95,82/99,26
Al	0,60	0,00/0,23	0,36/0,39	0,00/0,36	0,44/0,14	0,53/0,39
Ag	1,54	1,46/0,29	5,04/2,17	7,35/3,58	4,19/5,04	3,55/0,00

In Table 4.5, time 0 corresponds to the initial solution (leached liquor), and the abbreviations c.a. and s.a. correspond to the studied parameters, with agitation and without agitation, respectively. The deposition of metals was performed at constant current, while the voltage decreased throughout the process, which can be explained by the variation of metals that adhered to the cathode surface.

Because the solution is highly acidic, the oxidation of the anode, along with other parameters, occurred quickly. However, it can be said that the cathode used is efficient to remove both copper and the metals mentioned from aqueous solutions. Table 4.6 shows the mass amount of silver in cathode deposits at different times and constant current density.

Table 4.6 - Silver concentration in constant current.

T	Deposit in the cathode		Q.t.	M.t	Silver concentration	
	c.a.	s.a.			c.a.	s.a.
15	0,052	0,003	792	0,885	7,53x10-4	9,65x10-6
30	0,299	0,078	1,584	1,77	1,50x10-2	1,69x10-3
60	0,002	0,002	3,168	3,442	1,39x10-4	5,37x10-5
90	0,036	0,096	4,7521	5,312	1,50x10-3	4,80x10-3
120	0,166	0	6,336	7,082	5,89x10-3	0

Where T= time (min.), c.a.= with agitation (g), s.a.= without agitation (g), m.t.=theoretical mass, Q.t. = theoretical load (C).

Taking into account the theoretical mass calculated by Faraday's Law presented in Table 4.5, it can be seen that the cathode deposition in this process was low, however, sufficient to remove considerable concentrations of silver, especially in 30 minutes of reaction. Apparently, an increase in the reaction time after 30 minutes once again causes an increase in metal deposition in the cathode, as well as a greater removal of silver.

To verify the concentration of silver in the final solution, Atomic Absorption measurements were performed on all aliquots collected at the end of each experiment at all reaction times, 15 - 30 - 60 - 90 and 120 minutes, with and without agitation. The results of these analyses are presented in Figure 4.4.

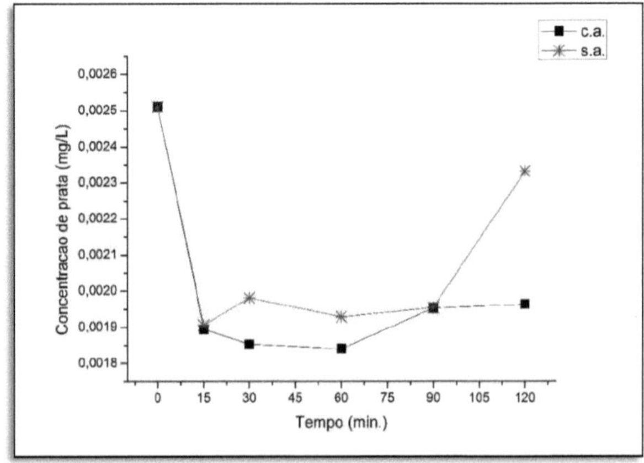

Figure 4.4 - Measures of silver concentration in solution, for processes with agitation and without agitation.

All aliquots taken every 10 minutes were analysed, however, concentrations of silver and other metals, at one time decreased, at another time increased to quantify the removal of silver at the end of each process, only the last aliquots were analysed. Starting with the initial solution, this assumes and in fact a decrease of the silver concentration in the solution occurred. The initial solution has a silver concentration of 0.00251 mg/L. The decrease in concentration already occurs in the first 15 minutes of reaction, and for the reactions with and without agitation the concentration of the metallic species is practically the same at this point, 0.00189 and 0.00191 mg/L respectively. In 30 minutes of reaction a higher concentration of silver in solution is observed for the process without agitation, which presents 0.00198 mg/L in the solution, since it presents higher concentrations of silver at the end of the reaction, for 30 minutes with agitation the concentration of 0.00185 mg/L.

With 60 minutes of reaction, the behavior is similar to 30 minutes, showing that the process without agitation, once again presents a higher concentration of Ag+ ion which is 0.00193 mg/L in relation to the agitation of 0.00184 mg/L. In 90 minutes, the concentrations are very close, presenting in the solution with agitation 0.00195 mg/L and 0.00196 for the reaction without agitation. A different behavior is observed for the 120 minutes of reaction time, where in the reaction without agitation, an increase in concentration is observed. This occurred because the electrode (anode) was completely oxidized in 100 minutes of reaction, with an unbalance of charges in the solution making it difficult to deposit the silver in the cathode. For the reaction with agitation, the behavior of the curve is maintained in relation to the other times, indicating a tendency to become constant. The concentration of silver in solution at this point is 0.00197 mg/L.

Analyzing these results, there is a better removal of the metal in 30 and 60 minutes, with agitation. This fact is important for time optimization. A time of 30 or 15 minutes could be considered in this case as well, since the silver concentrations are very approximate. The low removal of silver in this solution may be related to its reduction potential which is very close to copper, and a high reduction rate of copper in the cathode may be interfering with the deposition of silver as both compete for the same surface. The removal of copper in this solution will be performed in later stages of this study.

4.2.2. Morphology of deposited metals

The surface morphology of the products deposited and studied by Scanning Electron Microscopy - SEM coupled to a Dispersive Energy - EDS system are presented in Figure 4.5.

Figure 4.5 shows the morphologies of the deposits in the cathode for the studied deposition times and process with agitation.

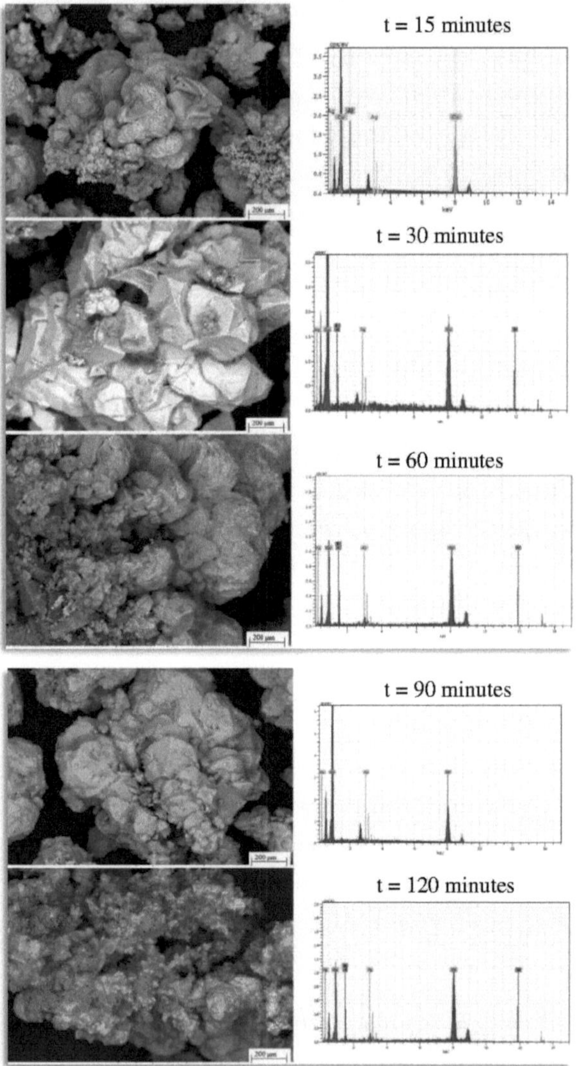

Figure 4.5 - SEM and EDS images of the cathode deposition products in 15, 30, 60, 90 and 120 minutes with constant density.

Due to the extremely acidic solutions, some factors are related to the simultaneous evolution of hydrogen, which impairs the electrochemical deposition process, such as the deposition of metals of interest in the cathode, resulting in brittle deposits and not adherent to the electrode due exactly to the evolution of hydrogen or secondary reactions still unknown in the ion environment of this solution. The morphologies of the species formed in the cathode are presented in Figure 4.5, and present structures formed on the surface of the electrodes as agglomerates of undefined forms, at all times studied. It is also possible to observe that even being a solution with several metals and of different concentrations, copper and silver can be deposited in the cathode while these two metallic ions coexist in the same solution. For the time of 15 and 120 minutes without agitation, no deposition in the cathode occurred, so it was not possible to perform SEM analysis. Figure 4.6 shows the morphology of the deposits in the cathode for the process without agitation.

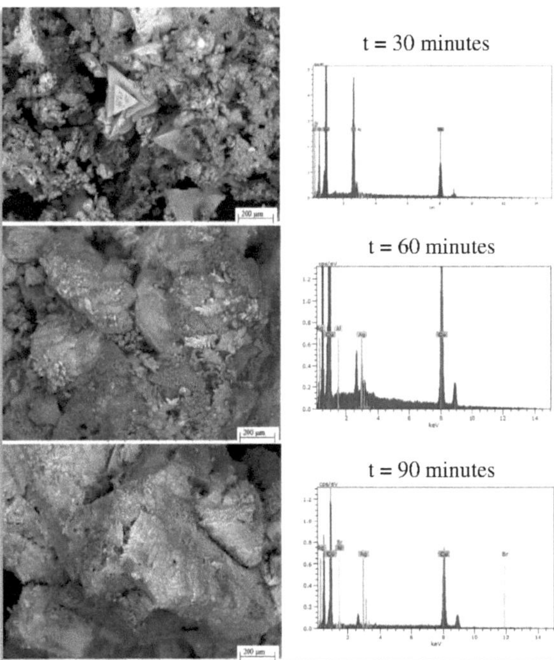

Figure 4.6 - SEM images of the deposition products in the cathode without agitation in 30, 60 and 90 minutes and constant density.

In the morphologies of the deposits for the process without agitation presented in Figure 4.6, the formation of metallic copper is observed in the reaction of 30 minutes, in the other structures, the beginning of the formation of a leaf is observed, in some parts of the deposits, which can also be attributed to the presence of silver.

The surface composition of cathode deposition products is further confirmed by the EDS mapping presented in Table 4.7, over time, with Cu, Ag and Al coexisting in the cathode.

Table 4.7 - Composition of the main metals present in cathode deposits.

	Time (minutes)				
Metal	**15 min.** **c. a/s. a.**	**30 minutes.** **c. a. /s. a.**	**60 min.** **c. a. /s. a.**	**90 minutes.** **c. a. /s. a.**	**120 min.** **c. a. /s. a.**
Cu (%)	95,93/0,00	83,93/96,45	93,1/94,04	99,29/92,56	95,52/0,00
Al (%)	3,03/0,00	11,47/2,20	5,14/5,21	0,00/2,99	4,40/0,00
Ag (%)	1,06/0,00	0,11/1,35	1,76/0,76	0,71/3,44	0,08/0,00

In Table 4.5, silver represents about 3.72% of the total deposition elements for the process with agitation, the proportion is increased to 5.55% for the process without agitation. Copper, at all times, presents the highest concentrations, and this occurs precisely because of the acid nature of the solution. These severe acid conditions dissolve the copper electrodes more quickly, generating a high amount of the same in solution. Even so, it can be seen that the electrodeposition of silver and copper can be achieved in this simple and easy to handle reactor, while the two metallic ions coexist in solution. In addition, there is the possibility to separate copper and silver by this same system.

4.3. Effect of the diluted solution in varying proportions on the electrodeposition of metals in the copper electrode

4.3.1. 1:3 dilution solution

Figure 4.7 shows the evolution of the process in the electrochemical cell in approximately 30 minutes of reaction to a dilution of 1:3, without mechanical agitation and 1.5 A of applied current.

Figure 4.7 - Electrochemical cell. (a) reaction in progress. (b) finished reaction.

In figure 4.7 (a) there is a difference in coloration, which evolves over the course of the reaction and is divided into three phases. The top colour is yellowish, the middle one remains light green, and the bottom one dark blue. The greenish solution, observed in the middle of the cell, between the electrodes, may be due to oxidation and at the same time the dissolution of copper ions in nitric acid. It also indicates that the copper ions are being reduced in the form of $Cu2+$. The yellowish green solution observed in the upper part of the solution is associated with the formation of $Cu0$ or $CuCl42-$ (Copper (II) Chloride) ions. The solution below the electrodes has a blue color due to the formation of $Cu\,(NO3)_2$ (copper nitrate). In the anode, even with the loss of mass, a thin white layer is deposited, which is probably associated with the formation of the chloride and oxide compound. A metal layer is deposited on the cathode. Figure 4.7 (b) shows the deposit of the cathode that fell into the solution, however it continued in its solid form, not dissolving in the solution, as was the case with the higher acidity solutions, used in the previous one.

The detachment of the tank when removing the electrodes from the solution is associated with the weight of the tank, as well as with the disconnection of the current, causing the tank to detach due to the absence of the force that kept them attached to the cathode. This white layer forming in the anode is observed soon in the first minutes of reaction, part of it falls in the solution as the reaction took place, and another remains adhered to the anode, even after the turning off of the source. At the end of the process, it was not possible to analyze it separately because it dissolves easily in the solution. However, the layer adhered to the anode could be analyzed by FRX and DRX. Figure 4.8 shows the electrodes at the end of the process.

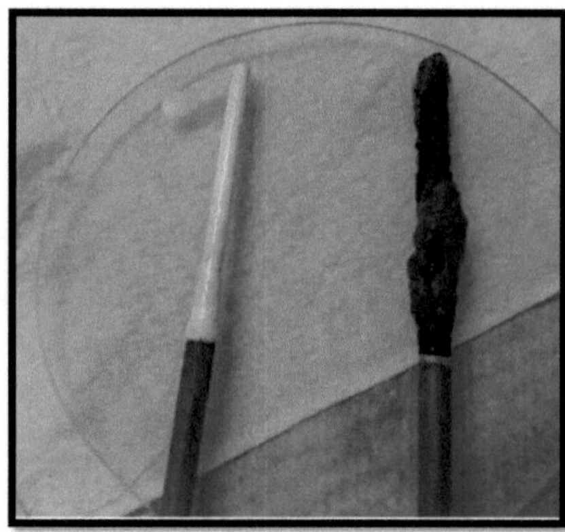

Figure 4.8 - Anode and cathode at the end of the reaction without agitation and dilution 1:3.

In the FRX analyses, concentrations of silver, sodium, chlorine, adhered to the copper anode were found. By these analyses, it is already possible to identify the formation of calcium chlorides, as well as copper and silver oxides, which have white coloration and are very soluble. Some of these compounds were found in DRX analysis.

All currents studied for 1:3 dilution and without agitation have the same behavior described above. Figure 4.9 shows the reaction to the process with mechanical agitation.

Figure 4.9 - Metal deposition with mechanical agitation.

In the reactions with agitation, it is observed that there was no formation of different colors in the solution, precisely because of the solution is in constant agitation. However, the deposition of metals in the cathode becomes more adherent, so that when switching off the system, it remains well attached to the electrode.

Also in Figure 4.9, the white layer is deposited on the anode, which is associated with sodium chloride, and remains adherent to the anode, even after the current is switched off. The cathode deposits, analyzed by FRX, are presented in Table 4.8, for the process with agitation, and in Table 4.9 for the process without agitation.

Table 4.8 - Composition of cathode deposits for the agitated process.

	0,5 A	1,0 A	1,5 A
Elements	%	%	%
Cu	94,458	94,099	88,397
Cl	3,435	4,25	9,079
At	0,301	0,190	1,593
Ca	...	...	0,331
Pb	0,194	0,194	0,103
Ag	3,191	1,398	1,571

The composition of the deposits in the cathode for the process with agitation, presents a high concentration of copper, followed by chlorine and silver. Although tin is in large quantities in the solution, no deposition rate has been verified in this process with agitation and all of it is still present in the solution. FRX analyses of the deposits were performed in an elemental way, however, compounds are also formed in the deposit as sodium chloride, $CaCl_2$, AgCl and PbO. Table 4.9 presents the composition of the deposit for the process without agitation.

Table 4.9 - Composition of the deposits in the cathode for the process without agitation.

Elements	0,5 A %	1,0 A %	1,5 A %
Cu	53,355	78,543	67,152
Cl	21,18	6,473	3,881
Pb	0,572	1,6	1,731
Al	0,768	0,309	0,183
Ni	0,101	...	...
Sn	12,552	11,993	26,288
At	6,165	2,165	1,435
Ca	1,042	...	...
Ag	0,391	0,99	0,764
Fe	0,274	...	...

In the composition of the deposits in the process without stirring high removal rates are obtained for tin, followed by copper, chlorine, silver, aluminum and lead. In addition, other metal fractions, as can be seen in the table, are adhered to the surface of the electrode, such as calcium, iron and nickel, and are only deposited in the 0.5 A current. The process without agitation favors the deposition of tin. In this deposit there are possible formations of compounds such as NaCl, Al_2O_3, PbO, $CaCl_2$ and SnO. The presence of these compounds is due to the fact that the cathode deposit has not been washed. Some elements such as nickel, iron and calcium may be in low concentrations and not detectable by FRX.

The aliquots collected during the process were analyzed by FRX. The results obtained for the process with agitation were not as satisfactory for the removal of the metals. It was observed that in all the aliquots collected, the concentrations of the metals in the solution did not seem to change, except for that of copper, which increased progressively as a function of the time and the current applied, precisely because of the oxidation of the anode, which was copper.

The increase of copper concentration in the 1:3 diluted solution is high, and is also due to its acidity, thus making the wear of the anodes faster, especially in the process with agitation. Parallel reactions also occur, making it difficult to remove the metals, among them, the hydrogen production. Even in the face of some parallel reactions, the copper deposit on the surface of the cathode for the analyzed currents are the main reactions, followed by silver, being the most probable elements of reduction in a solution of below zero pH. These results indicate and emphasize that copper is selectively reduced in the cathode by the applied process. As the reduction potentials of iron, lead, tin and other elements are smaller than those of the hydrogen ion, the electrodeposition of these metallic ions is more difficult to occur from a thermodynamic point of view. Therefore, the concentrations of these metallic ions in these conditions practically do not change with the reaction time.

In the process without agitation, there is a different behavior regarding the removal of tin in solution, high removal rates were achieved. What is probably occurring is the formation and release of hydrogen gas near the cathode region during the electrodeposition process, and the pH of the solution was increased. This high deposition of tin, especially in the reaction applying the 1.5 A current and without agitation, can also be explained by the tin ion formed in the solution. Sn^{4+} predominates in the reaction, reducing to the Sn^{2+} ion, Equation (17), and forming compound, probably with chlorine - Tin Chloride. Its reduction occurs following the reaction presented in Equation (18). Due to this increase in the pH of the solution, some of the tin ions were also precipitated during the reaction by the following equation.

$$Sn^{4+} + 2e^- \longrightarrow Sn^{2+} \qquad \text{Equation (17)}$$

$$Sn^{2+} + 2OH^- \longrightarrow Sn(OH)_2 \qquad \text{Equation (18)}$$

From this consideration, which in fact happens at this point, the activity of the tin ion is equal to the concentration of the tin ion, the pH calculated at the moment the tin hydroxide begins to form is 0.94 (Pourbaix, 1966).

In fact, tin hydroxides are precipitated, and at the end of the reaction, aliquots of this precipitate can be collected and it can be assessed whether the metallic tin, or tin compounds, were really being removed from the solution by electrodeposition. The aliquots collected at the end of the process and analyzed by FRX, show a decrease in the concentration of all the elements that were present in the solution at the beginning of the process, as shown in Table

4.8. Table 4.10 shows the concentration of metals at the end of each process, in all currents studied and without mechanical agitation.

Table 4.10 - Metal concentration at the end of each studied current and without mechanical agitation.

	Initial solution	0,5A	1,0 A	1,5^a
Elements	%	%	%	%
Cu	39,589	50,967	59,893	63,347
Cl	22,396	24,172	20,102	19,3
Sn	18,283	12,447	8,668	1,197
Ca	8,780	5,746	2,795	2,153
Pb	2,008	...	...	...
Fe	1,228	0,831	0,566	0,559
Ag	0,938	0,679	0,418	0,44
Al	0,749	0,688	0,599	0,373
Si	0,124	0,192	0,067	0,035
Ni	0,215	0,17	0,15	0,113
Cr	0,215	0,136	0,075	0,083
Ti	0,149	0,098	0,042	0,047
At	5,326	3,874	6,624	4,353

By chemical analysis, it can be affirmed that there is probable formation of sodium chloride, as well as silver chloride, because during the process, white powders that were being deposited in the anode fell in the solution, and were very soluble, characteristic of compounds formed by chlorides and oxides. At the end of the process, a considerable percentage of tin is still verified in all the applied currents, mainly for the 0.5 A current. This means that the current was low, minimizing the generation of ions for the solution as well as the transport of ions for the cathode. Tin removal rates for the studied currents of 0.5 - 1.0 and 1.5 A were 32%, 53% and 89% respectively. Lead is completely removed by deposition in the cathode, which is considered an excellent result due to its high toxicity in solution. It may be deposited in the cathode as an alloy Pb-Sn, in its oxide form or in its metallic form, even with its reduction potential being negative.

The aliquots collected, over time and analyzed by FRX, for the process without agitation and current of 1.5 A, showed excellent results, which was expected for other currents. Figure 4.10 presents the tin removal curve versus reaction time.

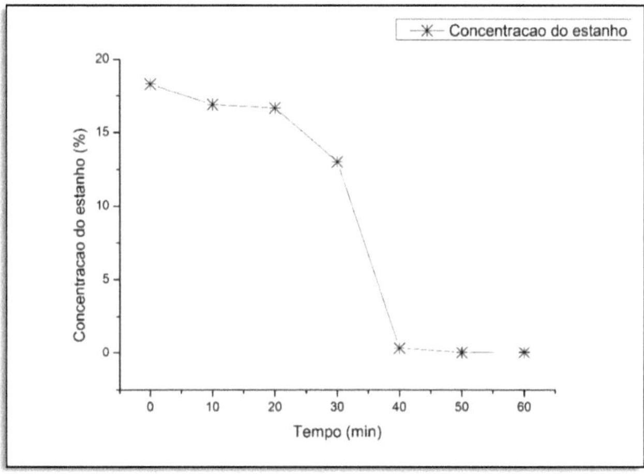

Figure 4.10 - Tin removal in electrochemical process without stirring and 1.5 A current.

The initial concentration of tin was 18.28%, and it decreases over time as it is deposited in the cathode. In 40 minutes of process, practically all tin is already deposited in the cathode, which would be a great time for its removal. It can be observed that at this point the curve follows the tendency to become constant, because practically all the tin was removed. The removal of tin in this process was 98%.

4.3.1.1. Atomic absorption

The results obtained in atomic absorption analysis are presented in Figure 4.11.

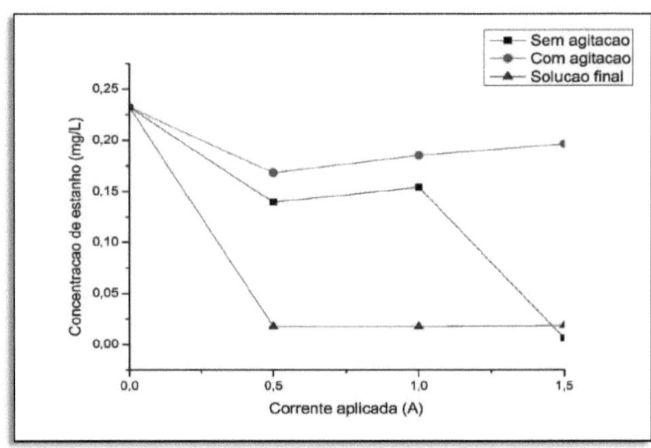

Figure 4.11 - Tin concentration at the end of each process - 1:3 dilution.

The analyses obtained by Atomic Absorption shown in Figure 4.11, present results similar to those found by FRX, where at the end of the process, there is in fact a decrease in the concentration of tin in the solution, and once again proves that the process without mechanical agitation favors the deposition and removal of tin, especially when the current applied to the process is 1.5 A.

4.3.1.2.   X-Ray Diffraction (XRD)

The deposits formed in the cathode and analyzed also by X-Ray Diffractometer, and is shown in Figure 4.12.

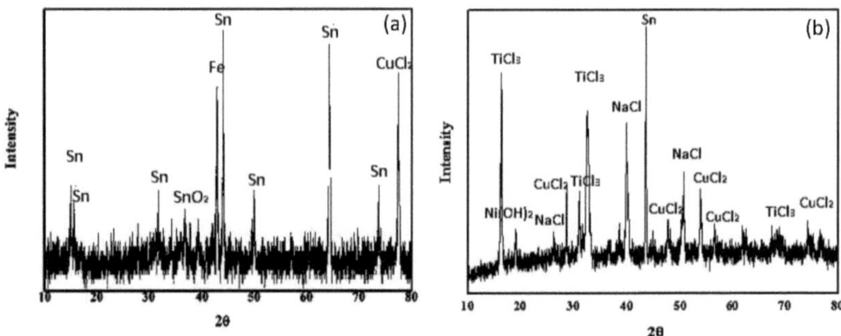

Figure 4.12 - X-ray diffractogram of the cathode deposit formed in 1.5 A current, with agitation (a), without agitation (b).

It is known that due to the heterogeneity of the sample, there is a variation of elements, and some elements are deposited in the form of compounds. According to the diffractograms a and b, in this sample copper and titanium are deposited as chloride, tin and iron appear in their metallic form. Another compound detected is tin dioxide, the presence of sodium chloride being proved in the second graph.

4.3.1.3.   Morphology of deposits -Dilution 1:3

The scraped copper cathode deposits characterized by Field Emission Scanning Electron Microscopy (SEM-FEG) are shown in Figures 4.13 and 4.14.

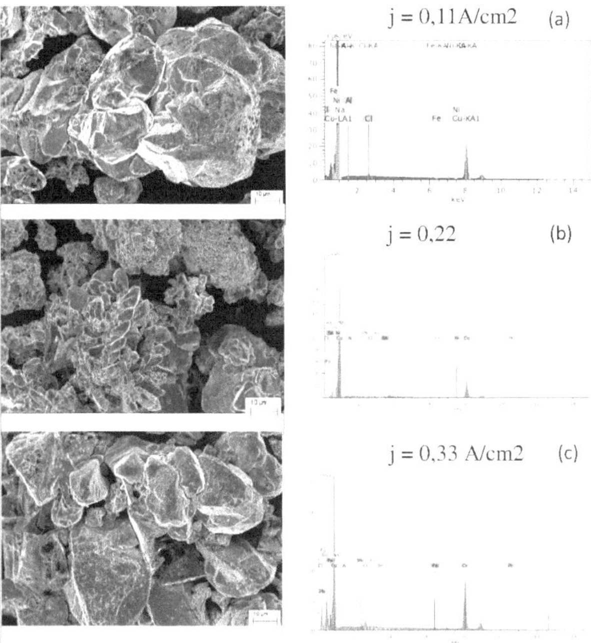

Figure 4.13 - Morphology of cathode deposits in current densities a) 0, 11 A/cm2 - b) 0.22 A/cm2 and c) 0.33 A/cm2, for the process with agitation.

The morphology of the deposits presented in Figure 4.13 shows similar characteristics according to the parameters studied, with agitation and without agitation. The structures of the metals deposited in the cathode, for the process with agitation, form flower-like agglomerates. This structure is more visible for currents of 1.0 and 1.5 A. The metal with the highest concentration in these deposits is copper, and in the whole area analyzed by SEM, the characteristics of agglomerated deposits remain, confirming that the morphology formed is due to copper. According to POPOV (2002), when a system involving metal deposition is characterized by a high exchange current (thermodynamic equilibrium where the liquid current is zero), spongy deposits are formed by using low cathode current density, while deposits of dendritic nature are formed by working with higher densities. Apparently, this is the behavior presented in morphologies.

Figure 4.14 presents the morphology of the deposits for the process without agitation, and follows the behavior of dentric deposits in higher currents, observed in the structure formed in current density of 0.33 A/cm2.

Figure 4.14 - Morphology of cathode deposits in current densities a) 0, 11 A/cm2 - b) 0.22 A/cm2 and c) 0.33 A/cm2, for the process without agitation.

For the process without agitation, the morphology is visibly interesting, formed by structures similar to leaves of a fern, in some points perfectly structured, as verified in the deposit of Figure 4.14 without agitation and current of 1.5 A. These deposits are mainly formed by copper and tin, forming a Sn-Cu alloy, among other metals, and are shown in Figure 4.15, which shows the mapping of the sample in leaf form. In an article published by Walsh, 2016, he discusses that in negative potentials, dendrites grow on tree-like branches. The same is presented in our study, proving that the structure is precisely the alloy formed between copper and tin.

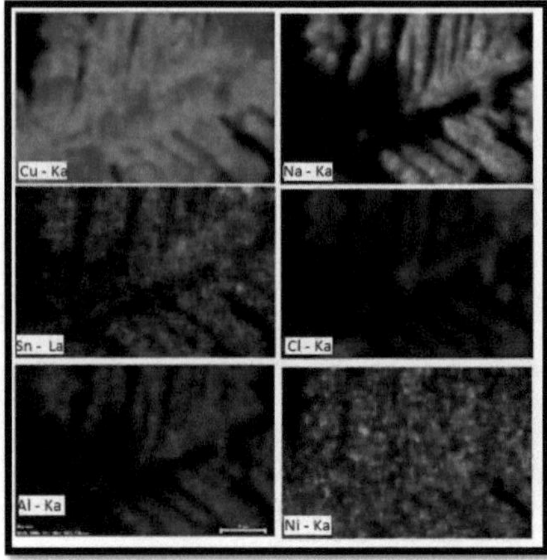

Figure 4.15 - Mapping of the structure of the deposit formed in the cathode - Without agitation and current of 1.5 A.

The morphology of metallic copper shows a more angular structure, shown in Figure 4.16, as seen along the foliage displayed by SEM. While a mixture of copper, tin, aluminum, nickel and sodium crystals results in the morphology of Figure 4.15.

Figure 4.16 - Angular structure of metallic copper - No stirring and 1.5 A current.

The surface composition of the cathode deposition products was quantified by EDS mapping and are presented in Table 4.11, with several metals coexisting in the cathode, for the process with and without agitation.

Table 4.11 - Composition of the main metals present in cathode deposits.

Metal	0,5 A c.a./s.a.	1,0 A c.a./s.a.	1,5 A c.a./s.a
Cu	97,16 / 30,43	96,47 / 79,52	97,03 / 84,90
Cl	0,44 / 50,09	0,45 / 15,04	1,4 / 9,32
At	0,96 / 6,88	1,64 / 2,30	0,75 / 2,49
Sn	0,00 / 3,86	0,00 / 1,71	0,00 / 1,42
Ni	0,94 / 0,83	0,87 / 0,52	0,00 / 1,25
Fe	0,32/1,24	0,25 / 0,13	0,57 / 0,51
Al	0,00 / 1,82	0,00 / 0,00	0,00 / 0,11
Pb	0,00/ 2,97	0,00 / 0,78	0,00 / 0,39
Ag	0,00 / 1,91	0,00 / 0,00	0,19 / 0,00

Of the base metals present in the cathode deposit, copper is the most deposited because of its high potential in relation to hydrogen. Besides copper, sodium, nickel, tin, zinc and precious metals can also be recovered in this way, as shown in Table 4.9.

4.3.2. 1:7 dilution solution

Figure 4.17 shows the evolution of the process in the electrochemical cell in approximately 30 minutes of reaction to a dilution of 1:7, without mechanical agitation and 1.5 A of applied current.

Figure 4.17 - Electrochemical cell. (a) reaction in progress. (b) finished reaction.

The behavior of the deposit in the cathode is repeated for the 1:7 diluted solution. At the end of the process part of the deposit falls into the solution. There was also a darker blue color in the solution below the electrodes, this solution was collected and analyzed by FRX for the three currents studied. The salt formed in the anode, in all processes, presents high concentrations, forming a kind of film around the active area of the anode, impairing the passage of current, which probably decreases the deposition of metals in the cathode. The evolution of gases, mainly hydrogen gas, is noticed as soon as the process is started, this is observed in both electrodes, mainly in the cathode, and it is visible to the eye in the process without agitation, presented in Figure 4.18.

Figure 4.18 - Formation of gases in the electrochemical process.

The formation of sodium chloride at the anode and the deposit in the cathode is shown in Figure 4.19, soon after removal from the solution.

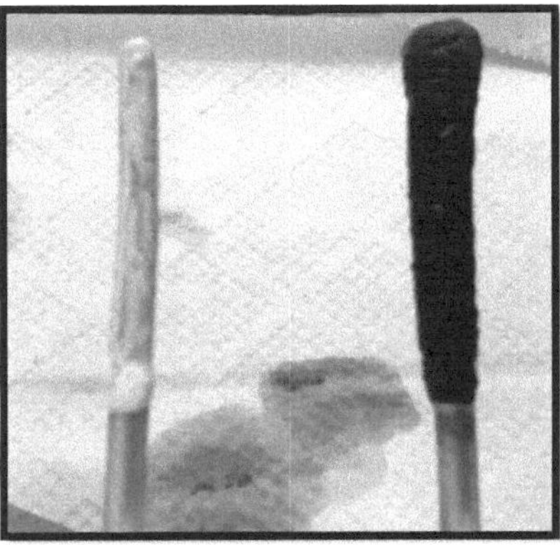

Figure 4.19 - Electrodes after the process is completed.

In the processes without mechanical agitation, it was observed that the deposit formed in the cathode, presented a darker coloration along the deposited area, and reddish at the tip. It is known that the reddish color is metallic copper, the rest of the deposit presents this color, probably due to the variety of metals deposited together with copper and tin, pricinperally the formation of silver oxide, which has black or brown color. Figure 4.20 illustrates the cathode deposition with mechanical agitation at 1:7 dilutions.

Figure 4.20 - Electrochemical process. (a) Electrochemical cell in process. (b) Electrode deposition at the end of the process.

Depositions at the anode are also verified in the processes with mechanical agitation within the first 10 minutes verified in Figure 4.20 (a). It is true that the formation of the film around the anode can impair the deposition of metals on the cathode, however, in all the currents studied in this process considerable removals of a variety of metals were obtained.

The cathode deposit shown in Figure 4.20 (b), has a reddish color in the whole deposited area, characteristic of metallic copper. One side of the electrode is thicker than the other side of the electrode, near the cell wall. This happens due to the effect of the current density, which shows to be more uniform between the two electrodes, with a greater deposit of metals occurring than between the electrode and the electrochemical cell wall. As previously mentioned, the deposit on the electrodes, when the process is with agitation, has a better adherence to the surface, making the deposit remain firm on the electrode, even after the current is switched off. Table 4.12 presents the composition of the cathode deposit for the process without stirring at 1:7 dilution in all currents studied.

Table 4.12 - Composition of the cathode deposit for the process without agitation.

Elements	0,5 A %	1,0 A %	1,5 A %
Cu	72,586	70,609	78,576
Cl	11,003	8,611	2,223
Sn	10,385	13,989	16,175
At	3,778	5,162	0,65
Ca	0,197	0,365	0,867
Pb	0,712	0,644	0,732
Ag	0,429	0,355	0,372
Al	0,018	...	...
Fe	0,836	0,198	0,163
Si	0,035	0,031	0,103
Ni	0,012	0,026	0.042
Cr	0,009	0,007	0,015
Ti	...	0,002	0,006

For solutions diluted in 1:7, the deposition behaviour was much more uniform for most metals compared to deposits at the 1:3 dilutions, especially for tin which shows an increasing increase in deposition rate as the current is increased. The highest deposition rates of tin were again for current density of 0.33 A/cm2, and in diluted solutions, different from silver, where its deposition is more efficient under extremely acidic conditions and at current density of 0.22 A/cm2. In diluted solutions, a small deposition of silver also occurs in the cathode. The composition of the deposits for the diluted solutions in 1:7, where the process took place with agitation, are shown in Table 4.13.

Table 4.13 - Composition of the cathode deposit for the process with agitation

Elements	0,5 A %	1,0 A %	1,5 A %
Cu	96,167	97,288	95,974
Cl	1,596	0,000	1,582
Sn	0,226	0,582	0,407
At	0,000	0,519	0,986
Ca	0,182	0,133	0,147
Pb	0,044	0,147	0,215
Ag	1,287	0,818	0,458
Al	0,227	0,000	0,000
Fe	0,000	0,130	0,133
Si	0,075	0,038	0,022
Ni	0,082	0,033	0,045
Cr	0,009	0,015	0,029
Ti	0,007	0,000	0,000

During the process of electrodeposition to current it remained constant. As the electrodeposition continues, there is a decrease in the concentration of ions in the solution as they are removed. Table 4.14 shows the concentration of metals in the solution at the end of each process, in all the currents studied and without mechanical agitation, which shows a decrease in the concentration of some elements as the current is increased.

Table 4.14 - Metal concentration at the end of the process without agitation.

| | | 0,5 A | 1,0 A | 1,5 A |
Elements	Initial solution	%	%	%
Cu	40,711	61,939	59,0955	61,2985
Cl	24,904	21,834	28,37	24,2705
At	3,638	1,91	5,3395	8,5405
Sn	17,265	6,717	2,766	2,789
Ca	7,326	4,126	2,4135	1,501
Ag	1,539	0,409	0,4305	0,441
Pb	1,966	0,997	0,416	0,416
Fe	1,195	0,867	0,51	0,375
Ni	0,209	0,176	0,082	0,06
Si	0,314	0,341	0,117	0,075
Cr	0,179	0,117	0,057	0,0415
Al	0,601	0,496	0,37	0,1695
Ti	0,154	0,071	0,033	0,0225

The process without mechanical agitation presents even more consistent results in 1:7 dilutions compared to 1:3 dilutions. Tin still presents considerable concentrations at the end of each reaction, much lower than the previous ones. The highest concentration present in the solution is for the applied current of 0.5 A. However, removal rates are 61%, 84% and 84% for currents of 0.5 - 1.0 and 1.5 A respectively. Nickel is practically all removed from the solution, accompanied by titanium, chromium, silicon, lead and calcium which have removal rates above 70%.

4.3.2.1. Atomic absorption

Atomic Absorption was performed in all final aliquots, that is, in the time of 60 minutes, in processes with and without agitation. The aliquots collected at the end of the process without stirring were also analyzed to verify the tin concentration, and are presented in Figure 4.21.

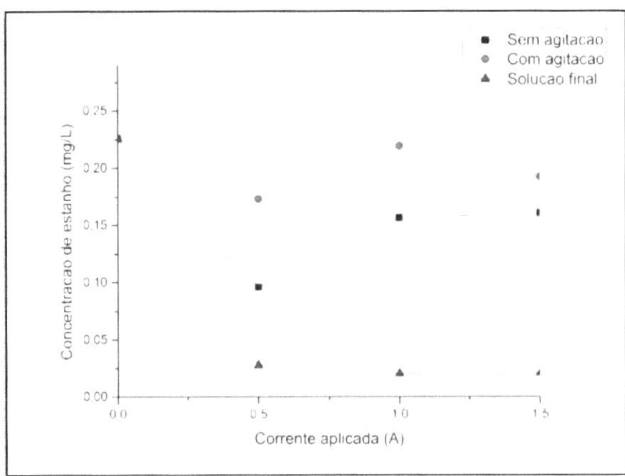

Figure 4.21 - Tin concentration at the end of each process - 1:7 dilution.

The Atomic Absorption analyses shown in Figure 4.21, show similar results to those found by FRX analysis, where at the end of the process, there is in fact a decrease in the concentration of tin in the solution, and once again proves that the process without mechanical agitation favors the deposition and removal of tin.

4.3.2.2. X-Ray Diffraction Analysis (XRD)

The DRX analysis was conducted in order to identify the compounds formed in the deposits, i.e., metallic, chloride or metallic oxides, confirming the deposition of the metals under study.

Figure 4.22 shows the DRX of a cathodic repository, in the process with and without mechanical agitation and 1.5 A current.

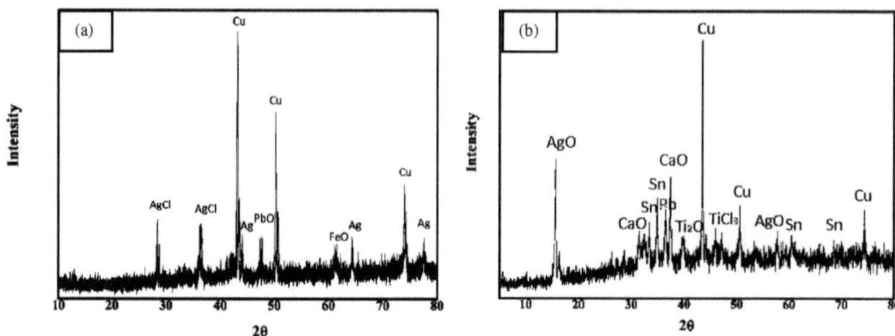

Figure 4.22 - X-ray diffractogram of the deposit formed in the cathode in a process, with agitation (a), without agitation (b).

The samples submitted to DRX analysis correspond to the processes with 1:7 dilution, with agitation, Figure 4.22 (a) and without agitation, Figure 4.22 (b) and 1.5 A current. The compounds found in the XRD analyses are in accordance with the results found by FRX. This proves the deposition of tin in its metallic form and silver oxide, as well as copper and other elements. Metallic silver, as well as silver chloride, appears in the deposits, where the process is carried out with agitation, proving once again the analyses by FRX.

4.3.2.3. Deposit morphology - 1:7 dilution

The scraped copper cathode deposits characterized by Field Emission Scanning Electron Microscopy (SEM-FEG) are shown in figure 4.23.

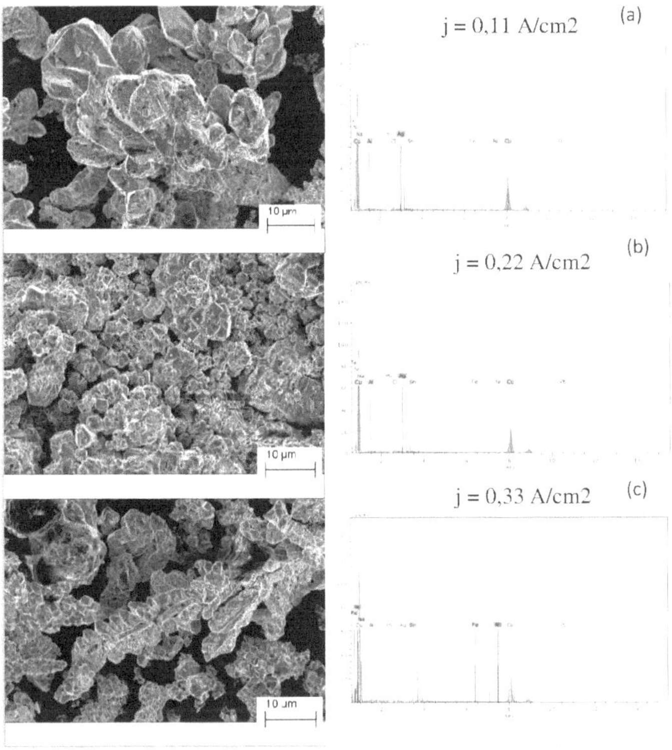

Figure 4.23 - Morphology of cathode deposits for the process with agitation in current densities a) 0.11 - b) 0.22 and c) 0.33 A/cm2.

The structures observed in Figure 4.23 for all deposits are in the form of numerous crystals with spherical sizes, stick shapes (Figure 4.25 - a) and regular squares. The spherical and square structures observed throughout the analysis and in the SEM images presented in Figure 4.24 correspond to crystals mainly CuO, as well as Cu0.

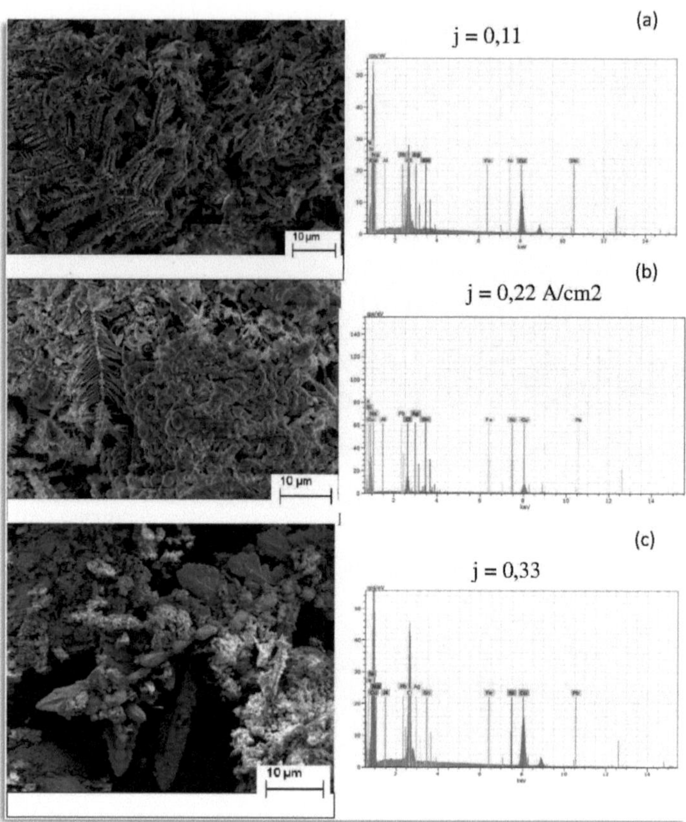

Figure 4.24 - Morphology of cathode deposits for the process without agitation in current densities a) 0.11 - b) 0.22 and c) 0.33 A/cm2.

In the morphologies of the processes, without mechanical agitation, the structure of the tin has not been identified, it may be forming a Cu-Sn alloy, as spherical formats, or a Pb-Sn alloy. The morphology of copper and its compounds was studied by Maroie, et al., 1984, confirming structures very similar to those obtained in this thesis.

At current densities of 0.33 A/cm2, there is a clear difference between the deposits, as the small grains formed continue to grow deliberately and thus new structures grow progressively. This behavior can also be attributed to the presence of silver, as mentioned by POPOV (2002) in their studies. The sticks observed in morphologies refer to lead, and are presented in Figure 4.25 (a), as well as the copper oxides with square cube-shaped structures presented in Figure 4.25 (b).

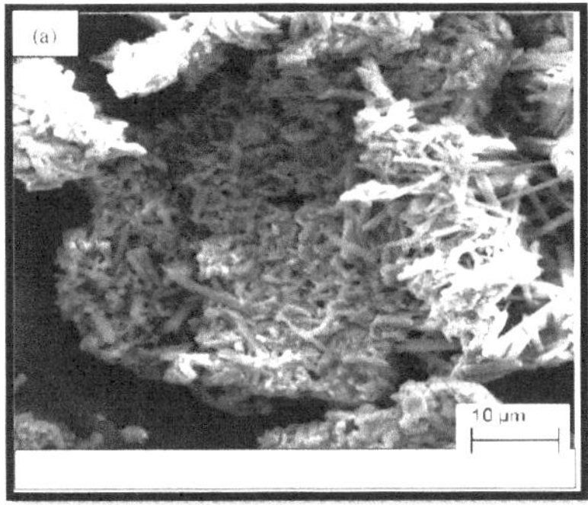

Figure 4.25 - Morphologies. (a) Metallic lead. (b) Copper oxide.

The structures, characteristics of copper, are formed according to each parameter studied. An example is metallic copper, which presents an angular structure when the process is maintained without agitation, verified in the morphologies of the deposits of the diluted solutions of 1:3, as well as 1:7. In the morphologies of the deposits in the process with agitation, copper is shown in a square structure, a cube, characteristic of copper oxide, as shown in Figure 4.25 (b). Similar structures to copper and lead were found in studies by Doulakas et al. (2000). They studied the recovery of copper, lead, cadmium and zinc from a synthetic solution by electrodeposition.

The surface composition of the cathode deposition products was quantified by mapping by EDS presented in Table 4.15, over time, several metals make up the cathode deposition.

Table 4.15 - Composition of the main metals present in cathode deposits

Metal	0,5 A c.a./s.a.	1,0 A c.a./s.a.	1,5 A c.a./s.a.
Cu	96,48/83,53	96,42/83,49	95,78/83,58
Ni	1,29/0,92	1,17/1,04	1,18/1,11
At	1,19/1,51	1,2/1,76	2,3/1,41
Cl	0,45/12,72	0,76/11,94	0,00/8,76
Fe	0,45/0,43	0,41/0,39	0,42/0,40
Al	0,14/0,00	0,00/0,00	0,20/0,00
Sn	0,00/0,78	0,00/1,25	0,00/4,69

Observing Table 4.15, copper always presents high cathodic deposition rates. We cannot but consider that there is in fact a refining of metals taking place. For IPPC (2000), when a copper anode, a metal used in the processes, containing impurities, and a cathode are placed in an electrolyte containing the metal in solution, the metallic ions are dissolved from the impure anode and pass through the solution to be deposited in the cathode. Copper, precious metals, lead and tin are refined in this way.

4.4. Variation of potential in the cathode

During all the electrodeposition reactions performed, a variation in the cathode potential was observed. The cathode potential, in all densities, tends to decrease in order to keep the current constant. At some moments of the process, the concentration of ions becomes low, and the processes of diffusion, migration or agitation, are no longer sufficient to maintain the constant current, being necessary adjustments in order to guarantee the current value, thus, the cathode potential will vary. In this case, the potential when moving to more negative values reaches the reduction potential of several elements, as shown in the previous tables of cathode deposition, causing an enrichment of elements in the deposits. For this reason, the results obtained show a variation of deposited metals, according to current density and cathode potential variation. Table 4.16 shows the initial and final cathode voltage in each process with diluted solutions, with and without agitation.

Table 4.16 - Initial and final cathodic voltage in each process with diluted solutions.

	1:3		1:7	
j (A/cm2)	c.a.	s.a.	c.a.	s.a.
0,11	0,9 - 0,7 V	1,2 - 1,0 V	2,2 - 1,5 V	2,4 - 1,8 V
0,22	1,5 - 1,2 V	2,3 - 1,7 V	2,7 - 2,5 V	2,3 -1,9 V
0,33	1,9 - 1,5 V	2,6 - 2,0 V	3,7 -3,2 V	3,2-2,9 V

Analyzing the data in Table 4.16, a decrease in cell voltage is observed in all processes. One can consider that there is a resistance between the anode and the cathode in this cell, so what occurs is that the current I went through this resistance causes a drop in potential, governed by Ohm's law. It is considered that there are two effects on the total change of resistance. It was observed that the removal of ions decreases the conductivity in the aqueous medium and the transfer of mass to the electrodes promotes an increase in its diameter and consequently its transverse area and this makes it possible to decrease the resistivity in the cell.

4.5. Faraday's Law

Using Faraday's law, it was then possible to obtain the respective tin masses for the two processes with dilution. Table 4.17 shows the mass deposited in the cathode for the two processes with dilution, which obtained the highest tin concentrations.

Table 4.17 - Tin concentration reached at different current densities.

	Deposit				Conc. Sn	
j = A/cm2	1:3	1:7	Q. t.	M.t.	1:3	1:7
0,11	0,26	0,0792	1,58	9,7x10-4	0,032	0,008
0,22	0,32	0,146	3,17	1,9x10-3	0,038	0,02
0,33	0,03	0,6071	4,75	2,9x10-3	0,0079	0,098

Q.t. = theoretical load (C), M.t. = theoretical mass (g), Sn concentration = tin concentration (g).

The theoretical mass, given by Faraday's Law, is not close to any of the experimental masses presented in Table 4.17. This does not mean that there was not a good deposition efficiency, because as already discussed in previous sections, the layer of metals deposited in the cathode, fall practically all in solution, leaving a small deposit in the cathode at the moment it was weighed, even so, the tin mass concentration in these deposits are higher than the calculated theoretical values.

4.6. Electrode loss and mass gain - Diluted solutions

The loss and mass gain in the electrodes were analyzed in relation to the applied current and are presented in Figure 4.26 for the 1:3 dilution. For all processes with diluted solutions, when the current used is low, the amount of gas formed during the process in the anode, mainly, is lower, which increases the useful life of the electrode, as it can be proved in the wear of the electrodes during the process presented in Figure 4.26 (a), relating the loss of mass versus applied current.

The faster and greater wear of the electrodes is associated with an anodic reaction of oxygen release, whose bubbles exert a mechanical force on the electrodes, thus releasing more ions into the solution. This behavior is observed in the process with mechanical agitation.

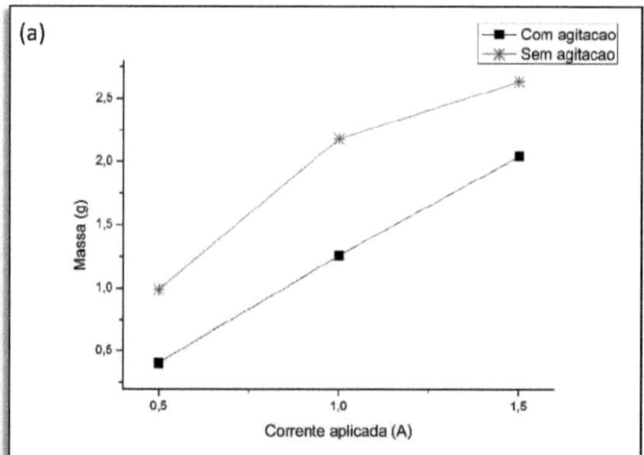

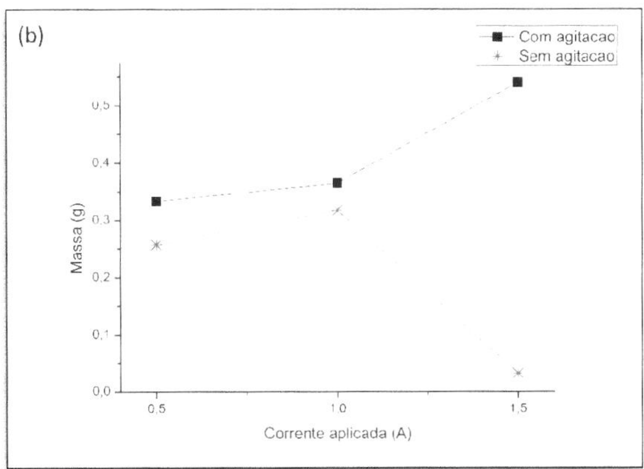

Figure 4.26 - Loss (a) and gain (b) of electrodes mass in relation to applied current.

The mass loss behavior occurs in a linear and increasing manner with the current applied. The mass gain in the cathodes, presented in Figure 4.26 (b), follows the linearity according to the current, except for the 1.5 A current and without agitation, where there is a sudden decrease of the mass gain. This occurred due to the fact that the deposit has fallen all in the solution when the process was finished. Table 4.18 shows the initial and final masses of the electrodes at each current density.

Table 4.18 - Electrode weight before and after each process.

Electrodes	P.A.			P.D.		
	j= 0,11	j= 0,22	j= 0,33	j= 0,11	j= 0,22	j= 0,33
Anode	23,4129	23,8208	23,427	23,0082	22,5619	21,386
Catodo	23,4627	23,2899	23,266	23,7199	23,6072	23,298

P.A. = weight before, P.D. = weight after, j = A/cm2

The mass losses and gains for the process with 1:7 diluted solutions show similar behavior to the previous ones, following the same mechanism of mass loss and gain, and are presented in Figure 4.27.

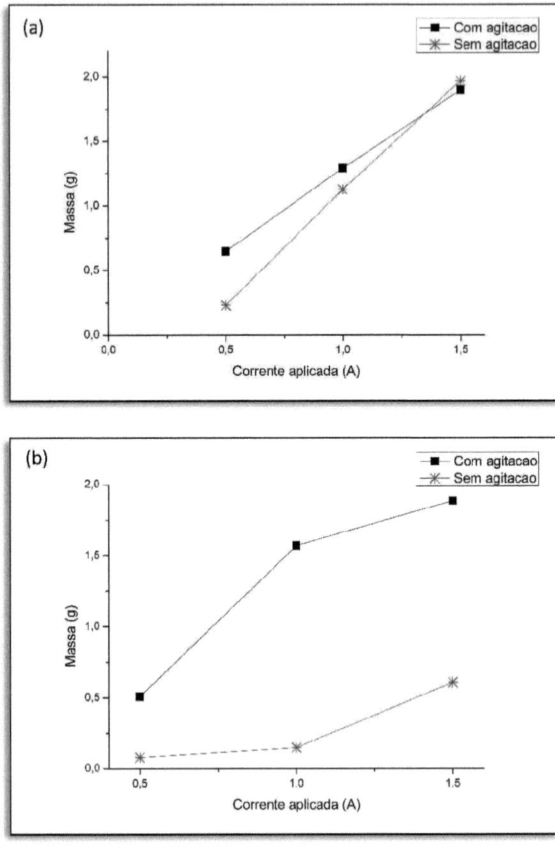

Figure 4.27 - Loss and gain of mass of the electrodes in relation to the applied current, loss of mass in the anode (a), gain of mass in the cathode (b).

The loss of mass is faster and greater for the process with agitation, precisely due to mechanical forces, mentioned above. Table 4.19 shows the initial and final masses of the electrodes at each current density.

Table 4.19 - Electrode weight before and after each process

Electrodes	P.A.			P.D.		
	j= 0,11	j= 0,22	j= 0,33	j= 0,11	j= 0,22	j= 0,33
Anode	24,1640	23,4767	23,6150	23,1752	21,2926	20,9793
Catodo	23,6258	23,2670	22,1494	23,9583	23,6321	22,6896

P.A. = weight before, P.D. = weight after

4.7. Increased active electrode area and different arrangements

In order to verify the deposition of metals in an electrode of different geometry, as well as in parallel electrode arrangements, some reactions were performed with the real solution, process with and without agitation and constant time of 60 minutes.

4.7.1. Active area in a circle for different current densities

The geometry of the active area of the electrodes was modified to verify the behavior of the current density in the extraction of metals, having as electrolytic medium the original solution. Once the reactions were carried out, it was verified that to increase the performance of the reaction in this electrochemical reactor and under these conditions, it is not enough to increase the area of the electrode, it is also necessary that the current density is equally distributed throughout the electrode so that the reaction occurs in a uniform manner. However, the current distribution is only uniform for very simple geometry electrodes, as occurred in the first arrangements. However, we can consider that in the circular geometry used, the current density is not uniform. Therefore, the current distribution will only be uniform if the electrode arrangement is geometrically similar. Figure 4.28 shows the mass of the cathodes in each reaction, with agitation and without agitation, and with current variation for 60 minutes, point 0, corresponds to the initial mass of the electrodes. The time remains constant for all.

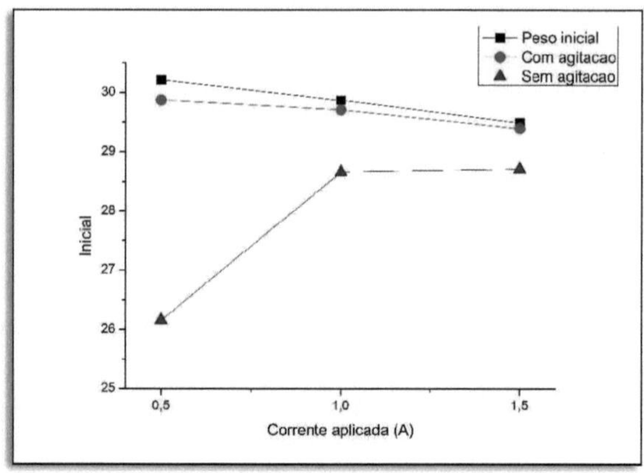

Figure 4.28 - Loss and gain of mass of the electrodes in 60 minutes of reaction with variation of current and active area.

The densities analyzed varied between 10, 20 and 30 A/cm2. It is important to say that there was no deposition of metals in the electrodes, what occurred and became evident in all processes was a loss of mass, which is more evident in densities of 10 A/cm2, evidenced in both processes, with agitation and without agitation, being more intense in the process without agitation. It is known that in this process, the oxidation of the anodes occurs, causing a loss of mass, and even with the increase in the loss of mass in the anodes, that is, the oxidation of the species Cu+/Cu+2 occurring, there is no influence on the transference of ions (deposits) to the cathodes.

4.7.2. Parallel arrangements for different current densities

Other reactions were performed with the original solution, where four electrodes were inserted in the electrochemical cell side by side, so that both pairs were used. A distance of approximately 3 cm between them was maintained. This bipolar arrangement was electrically connected to the power supply so that the same current was flowing through all the electrodes. Figure 4.29 presents the electrochemical cell with the parallel arrangement for the process with and without agitation.

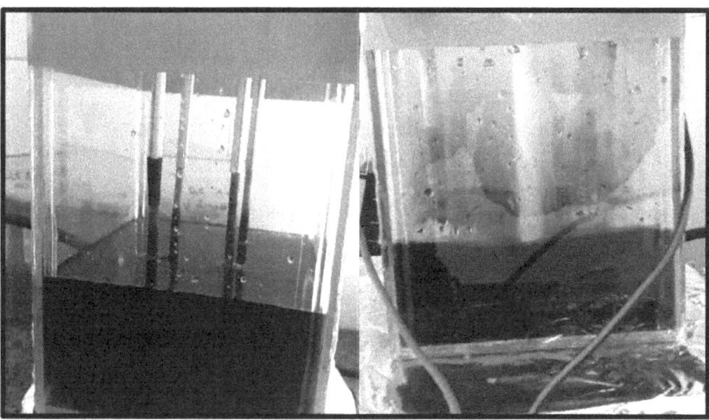

Figure 4.29 - Electrochemical cell with parallel arrangement for the process with and without agitation.

In this electrode arrangement, there was no metal deposition in any of the studied parameters, with or without agitation, and no current proved efficient. What happened in fact was the rapid wear of the electrodes, mainly in the process with agitation, and visually observed a formation of gas generated since the beginning of the process. The rapid wear of both electrodes makes the solution very dark, as soon as 10 minutes after the process, which makes it difficult to follow the deposition or not of the metals. In this same process, an increase in the temperature of the solution was verified. This may be due to the shock of the large amount of free electrons against atoms of the conductive material, transforming the electrical energy into thermal.

In fact, it should be noted that as there is variation in the active area of the electrodes, there is also a variation in the current density, that is, the current density is being increased in the two cases described above, so it is important to investigate which of the two parameters is making it difficult to deposit the metals in the cathode, or causing the loss of mass of both electrodes. In the present thesis, this parameter was not investigated.

Chapter V

Conclusions

5. Conclusions

The process developed in this study proved to be a viable alternative and allowed the recovery of copper, tin and silver from PCIs, through a relatively simple technology, without many environmental problems and also without many energy expenses related to other processes.

In the results for the leaching, the temperature favoured the dissolution of the metals in regia water, where the temperature of 80 °C and a time of 6 hours favoured a greater mass of dissolved metals. For copper 100% was obtained, for tin above 95% and for silver above 70%.

The parameters with and without agitation, in the real solution, indicated that in the process with agitation, the efficiency in the recovery of silver was significant at all times and currents studied, having a better deposition of the metal when in current density of 0.22 A/cm2. Removal rates were higher than 50%, proving to be a very important parameter in the recovery of silver and other metals from Printed Circuit Boards.

In diluted solutions, the processes without stirring favoured the removal of tin with removal rates above 70% and in some cases reaching approximately 100%.

The recovery of copper, tin and silver in the precipitate from leached liquors, Printed Circuit Boards (PCB), reached significant values in all parameters studied in this system.

Taking into account the experimental results obtained here, concentrates of tin, copper, lead, silver and other metals can be recovered from obsolete computer PCIs using a hydrometallurgical approach based on acid leaching, followed by electrochemical deposition of inorganic species in solution.

Scanning electron microscopy (SEM) and energy dispersion spectroscopy (EDS) analyses indicated characteristic structures of the deposited metals. The current applied, together with the processes with and without agitation, significantly influence the movement of charged particles that, by adhering to the surface of the electrode, determined its morphology.

There are many advantages when using this process, among them we have: the ions can be recovered in their metallic form and can then be reused, without the generation of residual sludge and in this case studied, without the need to add chemicals and the ease of controlling the process, since the control variable is the electric current and the consequent reduction in the need for labor.

Faced with this study, this becomes an alternative to relieve the bottleneck of economic development caused by the lack of resources, the recovery of metals from electrical and electronic waste, Printed Circuit Boards, has great development potential, especially because precious metals represent more than 70% of the value, with copper representing approximately 20%.

References

References

Arslan, F., Duby, P.F., 1997. Electro-oxidation of pyrite in sodium chloride solutions. Hydrometallurgy. V 46, p. 157 e 169.

Baldé, C.P., Wang, F., Kuehr, R., Huisman, J. (2015). The global e-waste monitor - 2014. United Nations University, IAS - SCYCLE, Bonn, Germany.

Bard, J. A. and Faulkner, L. R. Electrochemical Methods Fundamentals and Applications. Nova York, John Wiley and Sons, 2001. 833p.

BORIN, C.A. (1986). Electrochemical benchtop reactor for removing metal ions from industrial effluents. Master's dissertation - USP-São Paulo.

BRAZIL. Law No. 11.445: National Policy on Basic Sanitation - PNSB, January 5, 2007.

BRAZIL. Law No. 12,305: National Policy on Solid Waste - PNRS, of August 2, 2010.

Caporali, S., Marcantelli, P., Chiappe, C., Pomelli, C.S. Electrodeposition of transition metals from highly concentrated solutions of ionic liquids. Surface & Coatings Technology, V 264, p. 23-31, Fevereiro 2015.

Cayumil, R., Khanna, R., Ikram - Ul-Haq, M., R. Rajarao, A. Hill, V. Sahajwalla. Generation of copper rich metallic phases from waste printed circuit boards, Waste Management. V 34, p. 1783–1792, 2014.

Chen, J.P., Chang, S.Y., Hung, Y.T. "Electrolysis," physicochemical treatment processes L.K.P. Wang, Y.T. Hung, N.K. Shammas (Eds.), Handbook of Environmental Engineering Series, vol. 5A, Humana Press, Totowa, NJ (2005) (Chapter 10).

Chi, X., Streicher-Porte, M., Wang, M.Y., Reuter, M.A. Informal electronic waster ecycling: a sector review with special focus on China. Waste Management. V 31, p. 731e742, 2011.

Choi, M. S., Cho, K. S., Kim, D. S. & Kim, D. J. Microbial recovery of copper from printed circuit boards of waste computer by Acidithio bacillus ferrooxidans. Journal of environmental science and health. Part A, Toxic/hazardous substances & environmental engineering, V 39, p. 2973–2982, 2004.

Commission, E. (2000). Integrated Pollution Prevention and Control (IPPC) - Reference Document on Best Available Techniques in the Non Ferrous Metals Industries (p. 399).

Cortés-López C., Reyes Cruz V.E., Veloz Rodríguez M.A., Urbano Reyes G., Cobos Murcial J.A., Nava Montes de Oca J.L. Electrochemical Selective Leaching and Deposition of Ag, Au and Pt from electronic waste. International Journal of Electrochemical Science. V 12, 8198 – 8216, 2017.

Cui, J., Forssberg, E. Mechanical recycling of waste electric and electronic equipment: a review, Journal of Hazardous Materials, V 99, p. 243–263, 2003.

Cui, J., Zhang, L. Metallurgical recovery of metals from electronic waste: A review. Journal of Hazardous Materials, V 158, p. 228–256, 2008.

Cunha, T. H. R. Investigation of transport mechanisms in quasi-dimensional electrolytes, 2010. Master's dissertation - Universidade Federal de Viçosa, MG, 2010.

De Marco, I., Caballero, B.M., Chomón, M.J., Laresgoiti, M.F., Torres, A., Fernández, G., Arnaiz, S. Pyrolysis of electrical and electronic wastes. Journal of Analytical and Applied Pyrolysis, V 82, p. 179–183, 2008.

Doulakas, L., Novy, K., S. Stucki, S., Comninellis, Ch. Recovery of Cu, Pb, Cd and Zn from synthetic mixture by selective electrodeposition in chloride solution. Electrochimica Acta. V 46, p. 349-356, Novembro 2000.

Doulakas, L., Novy, K., S. Stucki, S., Comninellis, Ch. Recovery of Cu, Pb, Cd and Zn from synthetic mixture by selective electrodeposition in chloride solution. Electrochimica Acta. V 46, p. 349-356, Novembro 2000.

Duan, C., Wen, X., Shi, C., Zhao, Y., Wen, B., He, Y. Recovery of metals from waste printed circuit boards by a mechanical method using a water medium. Journal of Hazardous Materials, V166, p. 478–482, 2009.

Dutra, A. J.B., Guimarães, Y. F.,   Santos, I. D. Direct recovery of copper from printed circuit boards (PCBs) powder concentrate by a simultaneous electroleaching–electrodeposition process. Hydrometallurgy. $\underline{V}$ 149, P. 63-70, Outubro 2014.

Forgarasi, S., Imre – Lucaci, F., A. Egedy, A., A. Imre-Lucaci. A., Ilea, P. Eco-friendly copper recovery process from waste printed circuit boards using Fe^{3+}/Fe^{2+} redox system. Waste Management. V 40, p.136-143, Junho 2015.

Fornari, P. E., Abbruzzese, C. Copper and nickel selective recovery by electrowinning from electronic and galvanic industrial solutions. Hydrometallurgy. V 52, p. 209–222, 1999.

Frias, C., Lozano, J. I., Palma, J., Diaz, G. Valorisation of effluents from the printed circuit boards manufacturing industry. In: Rewas'2004 – Global Symposium on Recycling, Waste Treatment and Clean Technology. Madrid, Spain. p. 2635-2644, 2004.

Gamboa, S.A.; Sebastian, P.J.; Rivera, M.A. Characterization of p-CdTe obtained by CVTG tellurization of electrodeposited CdTe. Solar Energy Material and Solar Cells. V. 52, p. 293-299, Maio 1998.

García-Gabaldón, M., Pérez-Herranz, V., García-Antón, J., Guiñón, J. L. Electrochemical recovery of tin from the activating solutions of the electroless plating of polymers Galvanostatic operation. Separation and Purification Technology, V 51, p. 143–149, 2006.

Gluszczyszyn, A., Zakrzewski, J., Smieszek, Z., Anyszkiewicz, K. Secondary gold, recovery from electronic scrap in Poland. Recycling of Metalliferous Materials. IMM. The Institution of Mining and Metallurgy. p. 87−91. Birminghan, Inglaterra. Abril 1990.

Goosey, M., & Kellner, R. A Scoping Study End-of-Life Printed Circuit Boards. Environmental Working Group, (August), 2003.

Gouveia, A. R. Printed Circuit Board Metal Recovery by Hydrometallurgical. Integrated Master in Environmental Engineering - 2013/2014 - Faculty of Engineering, University of Porto, Porto, Portugal, 2014.

Guo, J., Guo, J., Cao, B., Tang, Y., Xu, Z. Manufacturing process of reproduction plate by nonmetallic materials reclaimed from pulverized printed wiring boards. Journal of Hazardous Materials. V163, p.1019 – 1025, 2009.

Hall, W.J., Williams, P.T. Separation and recovery of materials from scrap printed circuit boards. Resources Conservation Recycling. V 51, p. 691 e 709, 2007.

Havlik, T., Orac, D., Petranikova, M., Miskufova, A., Kukurugya, F., Takacova, Z. Leaching of copper and tin from used printed circuit boards after thermal treatment. Journal of Hazardous Materials. V 183, p. 866–873, 2010.

Hayes, P. C. Process Principles in Minerals and Materials Production. Hayes Publishing CO. p. 29. Brisbane, Australia, 1993.

Hoffmann J. E. Recovering precious metals from electronic scrap. JOM, V 44, p.43-48, julho, 1992.
Horkans, J. Effect of plating parameters on electrodeposited NiFe. Journal Electrochemical Society, V 128, p. 45-49, 1981.

http://www.mma.gov.br/política-de-resíduos-sólidos, access on 06/12/2018 at 15h.

https://exame.abril.com.br/brasil/brasil-gerou-15-milhao-de-toneladas-de-lixo-electronic-in-2016, access on 06/12/2018 at 14:57h.

Huang, K., Guo, J., Xu, Z. Recycling of waste printed circuit boards: A review of current technologies and treatment status in China. Journal of Hazardous Materials. V 164, p. 399–408, 2009.

Hwang, J.S. In Solder Paste for Electronics Packaging; Van Nostrand Reinhold; 1987.

Jandova, J., Stefanova, T., Niemczykova, R., 2000. Recovery of Cu- concentrates from waste galvanic copper sludges. Hydrometallurgy. V 57, p. 77 e 84.

Janssen, L. J. J., & Koene, L. (2002). The role of electrochemistry and electrochemical technology in environmental protection. *Chemical Engineering Journal, 85*(2-3), 137-146.

Juarez, C. M., Dutra, A. J. B. Gold Electrowinning From Thiourea Solutions. Minerals Engineering, V13, p. 1083-1096, 2000.

Kaya, M. Recovery of metals and non-metals from electronic waste by physical and chemical recycling processes. Waste Management. V 57, Pages 64-90, November 2016.

Kim, E. Y.; Kim, M. S.; Lee, J. C.; Jha, M. K.; Yoo, K.; Jeong, J. Effect of cuprous ions on Cu leaching in the recycling of waste PCBs, using electrogenerated chlorine in hydrochloric acid solution. Minerals Engineering. V 21, p. 121–128, 2008.

Kim, T.H.; Park, C.; Lee, J.; Shin, E. B.; & Kim, S. Pilot scale treatment of textile wastewater by combined process (fluidized biofilm process – chemical coagulation – electrochemical oxidation). Water Research, V 36, p. 3979-3988, 2002.

Kinoshita, T., Akita, S., Kobayashi, N., Nii, S., Kawaizumi, F., Takahashi, K. Metal recovery from non-mounted printed wiring boards via hydrometallurgical processing. Hydrometallurgy 69., p. 73-79, 2003.

Kirchner, J., Timmel, G., Schubert, G., 1999. Schubert, Comminution of metals in shredders with horizontally and vertically mounted rotors – microprocesses and parameters, Powder Technol. 105, 274-281.

Kongsricharoern, N., Polprasert, C. Chromium removal by a bipolar electro-chemical precipitation process. Water Science Technology. V 34, p. 109–116, 1996.

Kusakabe, K., Nishida, H., Morooka, S., Kato, Y. Simultaneous electrochemical removal of copper and chemical oxygen demand using a packed-bed electrode cell. J. Appl. Electrochem. V 16, p. 121–126, 1986.

La Marca, F., Massacci, P., Piga, L. Recovery of precious metals from spent electronic boards. In: Recycling and Waste Treatment in Mineral and Metal Processing: Technical and Economic Aspects, 16-20 Junho, Lulea, Suécia, 2002.

Lee, J-C., Jeong, J., Yoo, K., Yoo, J-M., Kim, W. Enrichment of the metallic components from waste printed circuit boards by a mechanical separation process using a stamp mill. Waste Management. V 29, p. 1132–1137, 2009.

Lee, M. S., Ahn, J. G., Ahn, J. W. Recovery of copper, tin and lead from the spent nitric etching solutions of printed circuit board and regeneration of the etching solution. Hydrometallurgy. V 70, p. 23-29, 2003.

Lekka M.; Masavetas I.; Benedetti A.V.; Moutsatsou A.; Fedrizzi L. Gold recovery from waste electrical and electronic equipment by electrodeposition: A feasibility study. Hydrometallurgy. V 157, 97–106, Outubro 2015.

Li X.Y.; Fan X.J.; Gu J.D.; Ding F.; Tong, A.S.F. Electrochemical wastewater disinfection: identification of its principal germicidal actions. Journal of Environmental Engineering **2004**, *130*, 1217–1221.

Luda, M.P. Recycling of printed circuit boards. Integrated Waste Management, V 2, p. 285-300, 2011.

M. Pourbaix, Atlas of Electrochemical Equilibria in Aqueous Solutions, Pergamon, Bristol,1966, p. 475-477.

Manetti, R. P., Chaves, A. P., Tenório, J. A. S. Recycling of metals from electronic scrap, in: Proceedings of 50th Annual Meeting of Associação Brasileira de Materiais e Metalurgia, São Pedro, Brazil, vol. 4, p 625-635, 1995.

Manetti, R. P., Chaves, A. P., Tenório, J. A. S., Obtenção de concentrados metálicos não ferrosos a partir de scrap electrónica (Non-ferrous metals recovery from electronic scrap), in: 51th Annual Meeting of Associação Brasileira de Materiais e Metalurgia, Porto Alegre, Brazil, V 4, p. 205-216, 1996.

Maroie, S., Haemers, G., Verbist, J.J. Surface oxidation of polycrystalline "alpha" (75% Cu et 25% Zn) and "bêta" (53% Cu et 47% Zn) brass as studied by XPS: influence of oxygen pressure. Applied Surface Science. V 17, p. 463–476, 1984.

Maroie, S., Haemers, G., Verbist, J.J. Surface oxidation of polycrystalline "alpha" (75% Cu et 25% Zn) and "bêta" (53% Cu et 47% Zn) brass as studied by XPS: influence of oxygen pressure. Applied Surface Science. V 17, p. 463–476, 1984.
Martinez-Huitle, C.A; Brillas, E. Decontamination of wastewaters containing synthetic organic dyes by electrochemical methods: A general review. Applied Catalysis b: Environmental, V 87, p. 105-145, abril 2009.

Martins, H.; Castro, L. A. Recovery of tin and copper by recycling of printed circuit boards from obsolete computers. Brazilian. Journal Chemical Engineering. V 26, no.4, Dezembro 2009.

Medeiros, N. M. *Characterization and Physical Separation of Obsolete Computer Printed Circuit Boards, 2015.* Master's Dissertation - UFRN - 2015.

Melo, R. A. C. *Study of Printed Circuit Board Leaching of UFRN Obsolete Desktop Computers*, 2017. Master's Dissertation - UFRN - 2017.

Ogunniyi, I.O., Vermaak, M.K.G., Groot, D.R., 2009. Chemical composition and liberation characterization of printed circuit board comminution fines for beneficiation investigations, Waste Mamage, 29, 2140-2146.

Oliveira, P. C. *Valorização de Placas de Circuito Imprestação por Hidrometalurgia.* 2012. 290 f. Thesis (Master) - Technical University of Lisbon. Lisbon, 2012.

Pant, D.; Joshi, D.; Upreti, M. K. & Kotnala, R. K. Chemical and biological extraction of metals present in E waste: A hybrid technology. Waste Management, V 32, p. 979–990, 2012.

Park, Y. J. E Fray, D. J. Recovery of high purity precious metals from printed circuit boards. Journal of Hazardous Materials. V 164, P. 1152–1158, 2009.

Parthasaradhy, N. V. (1989). Practical Electroplating Handbook. Prentice-Hall, New Jersey, 1989.

Petranikova, M., Orac, C., Miskufova, A., Havlik, T. Hydrometallurgical Treatment of Printed wiring boards from used Computers after Pyrolytic Treatment. In: European Metallurgical Conference. Innsbruck, Austria. July 2009.

Popov, K. I., DJOKIC, S. S., GRGUR, B. N. Fundamental Aspects of Electrometallurgy. Kluwer Academic/Plenum Publishers, New York, 2002.

POPOV, K. I., DJOKIC, S. S., GRGUR, B. N. Fundamental Aspects of Electrometallurgy. Kluwer Academic/Plenum Publishers, New York, 2002.

Pourbaix, M. Atlas of Electrochemical Equilibria in Aqueous Solutions, Pergamon, Bristol,1966, p. 475-477.

Pretorius R, Crouse P.L. Hattingh C.J. A multiphysics simulation of a fluorine electrolysis cell. South African Journal of Science; V 111, 5 pages, Agosto 2015.

Reyes-Cruz, V., Ponce-de-León, C., González, I., Oropeza, M.T. Electrochemical deposition of silver and gold from cyanide leaching solutions. Hydrometallurgy. V 65, p. 187-203, 2002.

Scott, K., 1995. Electrochemical processes for clean technology. Technical report. Royal Society Chemistry. Cambridge.

Scott, K., Chen, X., Atkinson, J. W., Todd, M., Armstrong, R, D. Electrochemical recycling of tin, lead and copper from stripping solution in the manufacture of circuit boards. Resources, Conservation and Recycling, V 20, p. 43-55, 1997.

Shokri, A., Pahlevani, F., Levick, K., Cole, I., Sahajwalla, V. Synthesis of copper-tin nanoparticles from old computer printed circuit boards - Journal of Cleaner Production, V 142, p. 2586 e 2592, 2017.

Silvas, F.P.C. *Use of hydrometallurgy and biohydrometallurgy for recycling of printed circuit boards*, 2014. PhD Thesis. Polytechnic School of the University of São Paulo. DEQ. São Paulo. 2014.

Simoni, M., Kuhn, E.P., Morfi, L.S., Kuending, R., Adam, F. Urban mining as a contribution to the resource strategy of the Canton of Zurich. Waste Management. V 45, p.10 e 21, 2015.

Sohaili, J., Muniyandi, S. K., Hassan, A. and Mohamad, S. S. Converting non-metallic printed circuit boards waste into a value added product. Journal of Environmental Health Science and Engineering. 2013.

Sum, E. Y. L. The recovery of metals from electronic scrap. JOM, 1991, p. 53-61.
Tenório, J. A. S., Menetti, R. P., Chaves, A. P. Production of non-ferrous metallic concentrates from electronic scrap. In: TMS Annual Meeting, Orlando, EUA, 1997, p.505–509.

Torabinejad, V., Rouhaghdam, A.S., Aliofkhazraei, M., Allahyarzadeh, M. Ni–Fe–Al2O3 electrodeposited nanocomposite coating with functionally graded microstructure Bull. Materials Science. V 39, p. 857-864, 2016.

Torabinejad, V., Rouhaghdam, A.S., Aliofkhazraei, M., Allahyarzadeh, M.. Ni–Fe–Al2O3 electrodeposited nanocomposite coating with functionally graded microstructure Bull. Materials Science. V 39, p. 857-864, 2016.

Veglio, F., Quaresima, R., Fornari, P., Ubaldini, S., 2003. Recovery of valuable metals from electronic and galvanic industrial wastes by leaching and electrowinning. Waste Manage. V 23, p. 245 e 252.

Veit, H. M. Pereira, C. C., Bernardes, A. M. Using mechanical processing in recycling printed wiring boards. JOM, V 54, p. 45–47, junho 2002.

Veit, H. M., Bernardes, A. M., Ferreira, J. Z., Tenório, J. A. S., Malfatti, C. F. Recovery of copper from printed circuit boards scraps by mechanical processing and electrometallurgy, Journal of Hazardous Materials, V 137, p. 1704-1709, Outubro 2006.

VEIT, H. M., *Copper Recycling of Printed Circuit Board Scrap*, 2005, 101f. PhD Thesis. Programa de Pós-Graduação em Engenharia de Minas, Metalurgia e de Materiais, Universidade do Rio Grande do Sul, Porto Alegre.

Vejnar, P., Hrabák, V. Recovery of non−ferrous and Precious Metals from secondary raw Materials. Recycling of Metalliferous Materials. IMM. The Institution of Mining and Metallurgy. Birminghan, Inglaterra. p. 275−281, Apr. 1990.

Volsky, A. & Sergievskaya, E. Theory of metallurgical processes. Moscow: Mir Publishers, 1978. p. 12–19.

Walsh, F.C., 2001. Electrochemical technology for environmental treatment and clean energy conversion. Pure Appl. Chem. V 73, p. 1819 e 1837.

Walsh, F.C.; Low, C. T. J. A review of developments in the electrodeposition of tin. Surface and Coatings Technology. V 288, p. 79-94, 2016.

Widmer, R., Krapf, H.O. E Kthetriwal. D. S., Schnellman, M.; Boni, H. Global perspective on e-waste. Environmental Impact assessment review. V 25, p. 436 – 458, 2005.

Xiu F. R. and Zhang F. S. Recovery of copper and lead from waste printed circuit boards by supercritical water oxidation combined with electrokinetic process. Journal of Hazardous Materials. V 165, p.1002-1007, Junho 2009.

Zhan, L., Xu, Z.M. State-of-the-art of recycling e-wastes by vacuum metallurgy separation. Environmental Science & Technology. V 48, P. 14092–14102, 2014.

Zhang, S., Forssberg, E. Mechanical recycling of electronics scrap - The current status and prospects. Waste Management and Research, V 16, 119–128, 1998.

Zhou, Y., Qiu, K. A new technology for recycling materials from waste printed circuit boards. J. Hazard. Mater. V 175, p. 823 e 828, Março 2010.

Zhou, Y., Wu, W., Qiu, K. Recovery of materials from waste printed circuit boards by vacuum pyrolysis and vacuum centrifugal separation. Waste Management. V 30, p. 2299-2304, Novembro 2010.

I want morebooks!

Buy your books fast and straightforward online - at one of world's fastest growing online book stores! Environmentally sound due to Print-on-Demand technologies.

Buy your books online at
www.morebooks.shop

Kaufen Sie Ihre Bücher schnell und unkompliziert online – auf einer der am schnellsten wachsenden Buchhandelsplattformen weltweit! Dank Print-On-Demand umwelt- und ressourcenschonend produziert.

Bücher schneller online kaufen
www.morebooks.shop

KS OmniScriptum Publishing
Brivibas gatve 197
LV-1039 Riga, Latvia
Telefax: +371 686 204 55

info@omniscriptum.com
www.omniscriptum.com

Printed by Books on Demand GmbH, Norderstedt / Germany